Rabbit Housing

*Planning, Building, and Equipping Facilities
for Humanely Raising Healthy Rabbits*

Bob Bennett

Storey Publishing

This book is dedicated to Scott and Alexia.

*The mission of Storey Publishing is to serve our customers by
publishing practical information that encourages
personal independence in harmony with the environment.*

Edited by Sarah Guare and Deborah Burns
Art direction and book design by Jessica Armstrong
Text production by Liseann Karandisecky and Jennifer Jepson Smith

Cover photography by © Lynn Stone: front, © Mark Kelley/Alaska Stock: back, and
   Marcelino Vilaubi/Storey Publishing: author
Interior photography by © Lynn Stone: 81 top right, middle right, and bottom left; 82–85;
   and 88; and © Mark Kelley/Alaska Stock: 86–87. Additional photography on page 81:
   courtesy of R. L. Wilson: top left, Dan Brink: middle left, Sarah Ouellette: middle center,
   and Debbie Vigue: bottom right.
Illustrations by © Steve Sanford

Indexed by Christine R. Lindemer, Boston Road Communications

Storey Publishing
210 MASS MoCA Way
North Adams, MA 01247
www.storey.com

Printed in the United States by Edwards Brothers Malloy
10 9 8 7 6 5 4 3 2 1

LIBRARY OF CONGRESS CATALOGING-IN-PUBLICATION DATA

Bennett, Bob.
   Rabbit housing / by Bob Bennett.
      p. cm.
   Includes index.
   ISBN 978-1-60342-966-5 (pbk. : alk. paper)
   1. Rabbits—Housing. I. Title.
SF453.8.B46 2012
616.02'73—dc23
                                               2012009376

Acknowledgments

A how-to book such as this takes *accumulated* know-how. What we have here is the sum of the knowledge, assistance, inspiration, and encouragement of a great many persons. They include my mother, Dorothy Bennett, and my brothers, Steve and Bill. S. E. Brand, my godfather, taught me some carpentry. Sonny Burke helped me make some of my first hutches, including one from half of a doghouse. Charlie Fox was a cop who made sure I put the doghouse back together.

Tony Pisanelli, Robert Noble, and Charlie Lyons taught me plenty, mostly by example, while I pestered them at their rabbitries. A girl named Joan was the only kid in my high school who did not tease me about raising rabbits, except for Joe Romano, who became a Catholic priest, and Frank Miglorie, now a college president. Joan didn't raise rabbits, but Joe and Frank did. Later, John Dack helped me build hutches.

Along the way my wife, Alice, endured my passion for rabbits. My children, Bob, John, and Alyssa, at first were enthusiastic about them and later tolerant.

Many rabbit raisers around the country, some of whom you will meet in this book and other books I have written about rabbits, provided lots of good ideas. I can't thank them enough. Dustin Bass and Geri Shepard of Bass Equipment Co. provided lots of photos and advice. So did Kevin Whaley of KW Cages. Christopher Igo let me use his computer when I left mine home.

Deborah Burns and Sarah Guare, my editors at Storey Publishing, and many others on the Storey staff, including Jessica Armstrong, Ilona Sherratt, and Mars Vilaubi, helped me pull this all together. All of the above, even if some didn't know it, contributed greatly to the finished product.

Bob Bennett
Shelburne, Vermont

"You can observe a lot by watching."

— Yogi Berra

Contents

Preface vi

CHAPTER ONE: **Planning Your Rabbitry** 1

A Brief History 1

Primary Considerations 10

CHAPTER TWO: **Building All-Wire Hutches** 15

Choosing Welded Wire 16

Sizing the Hutch 18

Planning the Door 19

    PLAN: *Easy Wire Hutch* 21

Multiple Hutches 30

    PLAN: *10 Small Hutches* 32

    PLAN: *Top-Opening Hutch* 35

CHAPTER THREE: **Outfitting the Hutch** 41

Types of Feeders 41

Watering Methods 47

Installing a Watering System 50

The Best Nest Box 56

    PLAN: *Wire Nest Box with Flanges* 58

    PLAN: *Wire Nest Box without Flanges* 61

Building Your Own Carrying Cage 64

CHAPTER FOUR: **Sheltering a Backyard Rabbitry**... 69

Simple Three-Wall Structures 71

Closed Buildings 90

Installing Hutches.................................... 99

Fencing All Around 104

CHAPTER FIVE: **Building and Outfitting a Big Operation**.............................. 109

Heating and Cooling................................. 109

Ventilation 110

Manure Management.................................. 113

CHAPTER SIX: **Tools, Equipment, and Routines**.... 119

The Necessities...................................... 119

Optional Items 122

Miscellaneous Equipment 124

Rabbitry Routine 126

Resources... 130

Index.. 132

Preface

I started housing rabbits in 1948 as a 12-year-old, with the objective of earning the Boy Scout merit badge for Rabbit Raising. My mother bankrolled me for a pair of pedigreed New Zealand Reds, and I began to raise them on Mendon Mountain in Vermont, a few miles from the foot of Pico Peak (now part of the Killington ski area, the largest complex of skiing mountains in the east).

Thousands of acres of the Green Mountain National Forest, populated with a broad spectrum of wildlife, separated the peaks from our backyard, an area of about eight acres. A small brook flowed 50 feet behind our house, and beyond it lay our woods, reached by a rustic log footbridge that I had built as a Boy Scout project. Perhaps because my mother had seen other outdoor rabbit hutches, she decreed that the Reds would reside across the brook and in those woods.

While my mother had come up with the money for the rabbits, it was up to me to provide a place for them to live. Somewhere I scavenged a wood packing box, about 2 × 3 feet, to which I nailed four legs made from maple saplings I cut in the woods. I made a door for the front of the box with scrap lumber and chicken wire left over from our henhouse, and I spread out some sawdust to make a nice "floor"

inside. The finished product looked a lot like my mother thought it would. To me, however, the new rabbit "hutch" looked pretty spectacular.

A breakthrough came when a neighbor offered me several abandoned wood-framed wire mink cages, which I soon remodeled for the rabbits, with tar paper roofs and hardware cloth floors supported by additional sapling legs. I made storm curtains out of burlap chicken-feed sacks. Before long I had a dozen hutches with a chicken wire fence all around to help keep out the many predators lurking in the forest. In due time I earned the coveted merit badge with the guidance of a counselor, Robert Noble, the same man who had sold me the Reds. While this rabbitry worked, its condition was testament to the resilience of rabbits. It also showed that my mother was perceptive. "Ramshackle," she called it. That was rabbitry number one, the first of seven I would build over the next 33 years.

Even today, when I see this sort of backyard eyesore, my mother's description comes to mind. And while you might get away with it in the woods on a mountain, you probably would not in suburbia. So I ask you this: Why build a ramshackle rabbitry when it's so easy to create something you'd like to show off?

Unlike chickens, you don't *keep* rabbits; you breed them. You produce more than you can keep, so you have to move some out. There are a number of ways to do that. You can keep some in your freezer until you are ready to put them in your pan or pot, or you can sell them as pets and breeding stock, as meat animals to processors and the public, or even as laboratory specimens. Selling rabbits often means you'll have visitors to your backyard. A great-looking rabbit operation will do more than almost anything else to enhance your business.

You can find a great many rabbit cages, hutches, and other equipment for sale in stores and on the Web, but often you can't tell which of the products are best to use. Frankly, I've been so appalled by much of what's available for sale that I decided to write this book. Here I share what I have learned from my own experience and from that of other rabbit raisers, to help you decide what works best for your own situation. I have included stories of rabbitries across the country, how they were built, and how they operate, in the hope that you find some of the ideas useful.

In this book we will examine not only the needs of the rabbits, but also of you, the caretaker, raiser, and breeder, and of the various constituents or stakeholders who might influence the enterprise. Taking the time to consider all the angles will go a long way toward ensuring the success of the operation. You, too, can house your rabbits safely, securely, and in good health, making the animals and yourself as comfortable as can be.

CHAPTER ONE

Planning Your Rabbitry

While this book is generally about housing and equipment for rabbits, it's really just as much about your backyard and the rest of your property. Most people who raise rabbits do so adjacent to their residence; if they lived on a farm they would probably be raising larger livestock. Certainly you don't need a ramshackle, smelly eyesore behind your house. You want to be as proud of the rabbits' quarters as you are of your own. To that end, we will develop an action plan for creating a successful rabbit operation that makes your home even more attractive and valuable than it is right now.

If only rabbits were as articulate as they are resilient, they could tell us exactly how to house them. But while they cannot utter any language, and indeed hardly ever make a sound, we can learn from their actions and reactions to their surroundings. I've been doing that for more than 60 years.

I have seen rabbits housed a great many ways — everything from boxes to dog houses and chicken coops. To their credit, they can survive in a lot of places, but you want more than mere survival. That's why this book is for you.

A Brief History

Years ago, in Europe and North America, most domestic rabbits ran loose in a barn built primarily for cattle, sheep, hogs, goats, horses, poultry, or all of these animals together. (Of course, in even earlier times the animals lived on the first floor of your house.) You wouldn't have to provide any accommodations for the rabbits because they burrowed into the dirt floors or under the hay or straw and formed a colony or warren. Feeding and watering was no problem, because you wouldn't bother with it.

The rabbits foraged for themselves, snitching hay and grain from the other species. That was the good news.

The disadvantages of the colony or warren were numerous. First, if you wanted a rabbit for dinner, you had to catch it. And you wouldn't know if it was a tender fryer or a stringy stewer until you took a bite. Second, you couldn't depend upon a set quantity of output or the timing of production because the rabbits made that determination themselves.

And there were bigger problems. Parasitic worms penetrated the animals from the dirt floors they shared with other species. That weakened the rabbits and left them thin, unthrifty consumers of the available feed and susceptible to debilitating or fatal disease.

Foxes, weasels, and other four-legged predators devoured some. Owls and hawks swooped down and flew off with others. Mature males often fought and injured each other. Some survivors hopped about outdoors only to absorb more parasites and provide more meals for wildlife or roaming cats and dogs.

In addition, because rabbits love to gnaw, they ate portions of the barn along with whatever else they could find. So you had a Swiss-cheese prison supervised not by the warden but by the inmates and their enemies.

The Morant

A somewhat better way to keep rabbits arrived in 1884. Designed and named by Major G. F. Morant, a British army officer, the morant was a portable, floorless outdoor enclosure for housing one or several rabbits of both sexes. The owner moved it around from place to place, while the rabbits took care of consuming, mowing, and fertilizing the grass. Because there was no floor, however, the parasites were happy and some of the rabbits burrowed underneath and escaped to the outside and an uncertain fate.

Improved versions included a wire mesh floor. The rabbits stayed inside but could still eat grass, which sometimes was supplemented by vegetable garden waste. In wet weather, it was dangerous. When snow covered the ground, it was useless.

You can find morants for sale today. Some are sold for poultry as "chicken tractors," but some are designated as rabbit homes. The new models are no better than the original version, and a plastic and wire edition sells for $650! You can only guess how hot the plastic enclosure part must be on a sunny day. If rabbits are anything, they're resilient, but those housed in morants will merely survive, not thrive.

✳ RABBITS ON THE GROUND ✳

In morants, as well as in warrens or colonies, the quality of the livestock and the quantity of their offspring were irregular at best. Mortality was high. A similar situation occurs today with rabbits kept in pens on the ground. When rabbits come in contact with the earth, they are susceptible to life-threatening parasites, as well as potentially wet conditions. Some people recommend doghouse-type structures with fenced "runs" attached. That can work for a pet — except during periods of snow, rain, and mud — but the problem of parasites is still there, and it won't help in a breeding project any more than a morant will.

A TYPICAL MORANT

On Rabbits Running Free

The idea that food animals should not be restricted in their movement seems to have broad appeal. One might easily assume that animals and birds running free are happier, healthier, and, ultimately, tastier. Many people assume rabbits should be raised similarly, rather than in some kind of confinement. While that approach may sound attractive, scientific research and practical experience prove otherwise.

Let's examine the supposed benefits of liberty.

Happiness

Do we know whether rabbits are happy or unhappy? If they are well-fed and comfortably housed, free from outside predators and inside bullies of their own species, disease-free, and even playful or affectionately responsive when handled, we might deduce that they are happy, if in fact this is an emotion they are able to experience. At the very least, they are at ease.

If they are not straining at their hutch doors to escape, one can conclude that they are not pining for long-lost freedom, chiefly because they have never possessed it. Well-bred domestic rabbits have been raised in hutches for thousands of (rabbit) generations and have never experienced freedom (except, of course, freedom from the environmental misery of disease and discomfort). In short, they don't miss what they never experienced — the freedom of adventurous and hazardous exploration.

In truth, if one of your rabbits escaped its hutch it wouldn't know where to turn, even if it wanted to flee. Should an escape occur in your absence, the vulnerable animal, which lacks the fear-inspired instincts of its wild cousins, becomes an easy mark for predators. Fortunately, some types of wire hutches are virtually escape-proof, and these are my primary recommendation. I want to keep my rabbits safe and secure. Only confinement will afford that kind of "happiness."

Health

Confined to the all-wire hutch, healthy rabbits will remain in top condition when properly fed and watered, because the hutch provides freedom from parasites and ensures ample ventilation while keeping the occupants safe from predators.

Animals that live on the ground are susceptible to worms. R. M. Lockley, a distinguished biologist and naturalist who undertook an intensive study of both pastured domestic and wild rabbits in the 1950s, found that several tapeworms infested the rabbits he studied. "Stomach, liver, genitals, skin and body cavity can be infected, resulting in large cysts or liver lesions," Lockley noted.

Most other farm animals are wormed either periodically or continuously with feed that contains a wormer. In all of my years of raising rabbits I have not needed to worm one, because they live in all-wire hutches and never come in contact with the ground.

Lockley also noted that all of his pastured rabbits, despite a dusting with insecticide, carried fleas on the ears and head, probably from burrowing in the earth. His rabbits also had a great incidence of coccidiosis, a debilitating intestinal disease caused by a parasite (*Eimeria steidae*) that leads to diarrhea and often death, especially in the very young rabbits.

Rabbits will survive on pasture — at least some of them will. We know that wild rabbits eat grass, other plants, tree bark, and roots. We also know that nature has equipped rabbits with a knack for producing many litters in quick succession in the springtime. It's a counterbalance to the fiercely high mortality rate of their offspring. Rabbits are universally applauded as paragons of reproductive perfection. They really need to be good at making more of their own because they are so vulnerable. Long experience with wire hutches proves they keep rabbits healthy — and less prone to disease than other species that are raised in contact with the ground.

Flavor

We know domestic rabbits can *survive* on grass, hay, and greens such as lettuce, but this results in slow growth. Many tons of rabbit meat find their way to the United States each year from China, where domestic rabbits eat greens almost exclusively, though the taste of this meat is inferior to that which comes from rabbits fed a grain-based diet. Farm-raised rabbits in the United States that are fed a balanced grain-forage pelleted ration in confinement produce a higher-quality carcass and much better flavor.

It's also clear that meat from our domestic rabbit, which is fine-grained, pearly white, and tender, is vastly superior in taste to wild rabbits' often stringy and well-muscled meat that is the result of the rabbits "running free." Just ask any rabbit hunter who has tasted the U.S. domestic product to render a comparison. Or, stop in at a French restaurant, where the hutch-raised *Lapin* entrées fetch upwards of $27 (and that's à la carte).

The Wood-and-Wire Hutch

Years ago, handy carpenters built wood-and-wire hutches with multiple compartments, sometimes two or more tiers high. These were widely used because neither chicken wire nor hardware cloth had any structural strength. This gets the rabbits off the ground, but this type of hutch also has some big flaws. Rabbits are death on wood. They gnaw. They spray urine. They shed their fur. They usually deposit their droppings in one corner of their enclosure. Most wood-and-wire hutches have wood supports under the wire, which create a manure buildup. Wood-and-wire structures also provide inadequate protection from rain, snow, and predators.

You can find a lot of wood-and-wire hutches for sale today, as well as plans from university agricultural extension departments, but I advise against them. The most expensive models, costing hundreds of dollars, resemble tiny barns or gazebos. The worst of them have wood supports under a wire floor, but they all have wood legs that are raring to rot. You can make modifications to improve them, however (see box on pages 8 and 9).

Note the dropping pan beneath and wire guard strips within this modified wood-and-wire hutch. See details on pages 8 and 9.

Useful Rabbit Tractors

Here's one way an outdoor wood-and-wire hutch can be beneficial: A friend of mine in New Jersey, Warren Burrows, who has a shed full of all-wire hutches, also has two wood-and-wire hutches on legs (with no wood framing members under the floor wire). A passionate vegetable gardener, he sets these hutches right in the middle of his cucumber patch during the summer. Young, growing does in each hutch provide a steady supply of manure for the plants, straight from the source.

You might call the structures "rabbit tractors," or even pseudo-morants. They do the same job but keep the rabbits safely off the ground (and occasional mud, of course). In the winter he moves the occupants he's keeping back to the shed. Warren grows a great garden.

The Best Option: The All-Wire Hutch

With the widespread availability of welded wire mesh came the all-wire hutch. The welded wire mesh permits the construction of single or multiple all-wire hutches without the need for any wood (or nails or screws), for a lot less money than structures requiring lumber. This type of hutch provides all the sanitary needs as well as essential protection and ventilation.

The all-wire hutch houses rabbits under the cleanest conditions of any livestock. Your prized rabbits never sit in manure or contract parasitic worms, and they receive plenty of ventilation. Sanitation and ventilation are of the utmost importance; without adequate provision for both, rabbits experience all kinds of health problems. The wire hutch is the only type of housing that meets these needs. What's more, with individual wire hutches you control the individual's diet, the selection and timing of matings, herd size, and heritability. I'll discuss the all-wire hutch in greater detail in the next chapter.

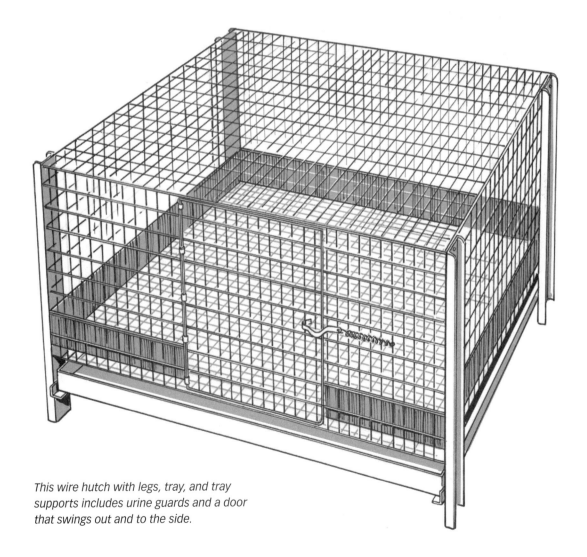

This wire hutch with legs, tray, and tray supports includes urine guards and a door that swings out and to the side.

Modifying a Wood-and-Wire Hutch

If you have already built or bought a wood-and-wire hutch, you probably know the drawbacks: the gnawing of the wood framework; the pile of manure in the corner; the attacks from below by wild animals or the neighbor's cat; and the assault by wind, rain, and snow. Here are some ways to renovate it to help solve those problems.

Protect the Wood

Rabbits love to chew the wood framework, particularly the door frame, and any exposed corner they can get their teeth around is fair fodder. Keep their choppers at bay by screwing pieces of L-shaped aluminum or steel angle to the edges of the wood framing members. L-angle is available at home centers and hardware stores in various lengths. Cut the pieces to fit with a hacksaw, and drill holes for screws.

Divert the Waste

Because many of these hutches inexplicably have wood support members under the perimeter of the floor instead of above, the manure sticks there, especially in the corners. To solve this problem, first clean off the corners with a wire brush, then screw aluminum flashing all around, with the flashing bent inward to deflect droppings and urine so they fall unobstructed to the floor. Cut pieces of flashing about 4" wide and bend them in half. The lower 2" half will clear the ¾" or 1½" framing member with room to spare. Screw the flashing to the wood. If the front is wire mesh, either wire or hog-ring the flashing to the mesh.

Another option is to purchase galvanized urine guards from an equipment supplier, such as Bass Equipment Company (see Resources).

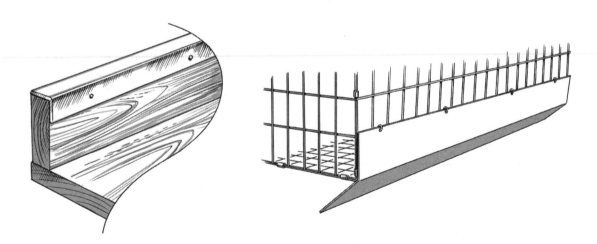

A metal strip protects the wooden edge of a wood-and-wire hutch (left). A urine guard keeps urine (and manure) directed toward the wire floor and away from wood support members (right).

Add Trays

If the hutch floor is only a foot or two above the ground, you can screw L-shaped metal angle sections on each end, from front to back, in the manner of drawer slides. Then add a galvanized or plastic tray on top of the angle. The tray not only catches the manure but also serves as a protective barrier against animals and drafts coming from underneath.

Prefab trays are available from rabbit equipment suppliers, or you can easily form one from a sheet of galvanized steel, available at home centers and hardware stores. Simply cut into two opposite edges about 2" deep and 2" from each corner, then fold up all four edges of the sheet. Fold over the extended ends of the uncut sides like tabs, and fasten each corner with a sheet metal screw. You might also bend the edges over about ½" and hammer them down so they won't be sharp (do this before folding up the sides).

Shield against the Elements

If rain, snow, or hot sun can reach into the hutch, staple an "awning" of canvas, plastic, or gardener's shade cloth along the top edge over the exposed face of the hutch. A plywood section added to the front of the roof to provide an overhang can also help. If deep snow is common, add height to the hutch by screwing 1×3 lumber leg extensions at the four corners. To ensure stability of a higher structure, you could insert these legs into concrete foundation blocks, nail them to railroad ties or landscape timbers, or bury them in the ground a foot or so. You can also place the hutch in a shed, garage, or other outbuilding. Do not put it in the house — a bad idea for you and your rabbit; if you try it, before long you will know what I mean.

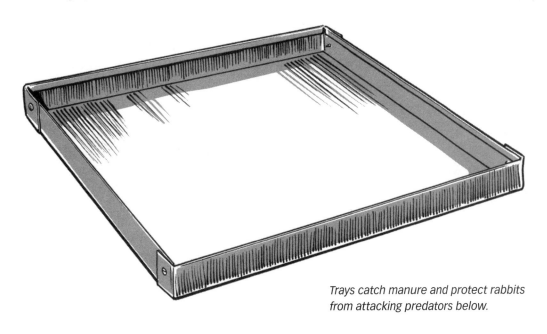

Trays catch manure and protect rabbits from attacking predators below.

Primary Considerations

As the president and CEO of your rabbitry, your first responsibility is the same as it is for the leader of a country: security. Your job is to provide a secure rabbitry that protects your animals from disease, parasites, predatory wild and domestic animals, and even human threats to well-being and success. Of course, you also need to safeguard your rabbits from themselves, which requires individual living arrangements for some and group homes for others. For example, to prevent fighting and possible injury, each mature male will need its own space, while growing does can be housed as a group for the early months of their lives.

To start the planning process, ask yourself these questions:

- *How many rabbits will I have, now and in the future?*
- *What size should the hutches be?*
- *What type of shelter will be around the hutches?*
- *Where will I locate the housing?*
- *How will I maintain a rabbit population consistent with the space I provide?*

Size

The size of your rabbitry will depend on how many rabbits you want to raise. Like most people starting out, you will probably begin with a small number of rabbits, but you ought to have an ultimate number of rabbits in mind before you start to put together your rabbitry. Each adult doe that will be producing a litter needs nearly a square foot of floor space for each pound she weighs; we'll discuss the size of hutches in more depth in chapter 2. What's more, the kind of rabbits you raise will determine rabbitry size because rabbits range from 2-pound dwarfs to 20-pound giants.

To help determine your rabbitry size, consider what type of operation you want: A few rabbits for pets or 4-H shows? Some for the family table, with perhaps breeding stock for sale? Angoras for their wool? Or maybe some meat rabbits for direct sales to friends or even processors, restaurants, and stores?

I learned that the kind of rabbits you raise should be the same ones you can consume or that somebody else might want, because you can't keep all of them, even if you wish you could. The number of rabbits you raise should be in line with the demand. As a beginner, it's hard to know how many to produce because you don't know how many you can sell. Even though you can consume the product yourself if you can't sell it (nice insurance, indeed), there's a limit to that, too.

Two important tips:

- *Line up some customers before you start producing, and remember that you can't sell what you don't have.*
- *Pay attention to inventory control.*

Your incentive to sell correlates directly with how many full hutches you have and how many more litters are on the way with no place to put them.

Weather

Your rabbits need protection from the elements. Will the hutches be housed simply with a roof and a little bit of siding, or in a closed building? You can place the hutches under a simple outdoor structure that might resemble a table with a pitched top or roof, inside an existing building such as a garage or tool shed that has been modified for rabbits, or a rabbitry structure that you either build from scratch or remodel from any of the many existing types of outbuildings you can buy. When choosing your structure, consider your resources (closed buildings may require more time and money to make them rabbit-safe); the needs of your rabbits (they will need good ventilation and special features); your own comfort (visiting your rabbits in a storm isn't much fun if there isn't a roof over your head as well); and the overall look of your property.

You'll need to visit your rabbits every day, so you should locate the building near your house. Later I'll discuss a lean-to style rabbitry that goes on the side of your garage.

Neighbors and Laws

Consider your family members and neighbors. Your rabbitry must meet the neighbors' approval (or be unknown to them — a distinct possibility). In addition, your municipality may impose zoning regulations that could imperil your plans.

Many municipalities have no laws specifically about rabbits. In some areas, laws

Decoy Rabbit

I know a man who raises quite a few rabbits in all-wire hutches in what his neighbors believe to be a tool shed. He also keeps a wood-and-wire hutch in his backyard with one rabbit in it. He calls this a *decoy*. The reason? There's an ordinance in his town against having several rabbits, but one pet is okay with the municipality, and the neighbors think that's all he has. That hutch always stays sanitary, attractive, and odor-free!

may be ambiguous, not mentioning rabbits by name but merely discussing "farm animals." Others may have ordinances that are easily circumvented. The fact remains, however, that even where there are no rules against rabbits, if your rabbitry doesn't pass the sniff test of family and neighbors, it won't get off the ground, or, even if it does, it won't last long. Where specific ordinances are in effect, you can bet they were passed because one or more rabbit raisers were operating with a ramshackle, smelly eyesore in their backyard. An attractive or unobtrusive rabbitry stands the best chance of success.

Having the knowledge to build a successful rabbitry, regardless of the ordinances, is vital because such laws can always become more strict, no matter how permissive they are at present.

Lops of Love at Home — *in New England*

Blackberry Farm
Debbie Vigue

LOCATION: Northern New England

AVERAGE RABBIT POPULATION: 50

GOAL: Show

BREED: Holland Lops

BEFORE I TELL YOU about her rabbitry, I want you to know about my friend Debbie Vigue, who lives in northern New England.

She is a constant source of ideas that other rabbit raisers never seem to think of. For example, when nobody wanted to be a writer at area rabbit shows — a tedious volunteer job entailing recording judges' oral remarks, hour after hour, for exhibitors during judging — Debbie came up with a special raffle just for them. The winner takes home a coveted basket of Maine lobsters or other sought-after items. Now people vie for the job. Here's another: She passed along a recipe for using the "fines" at the bottom of the bag of rabbit pellets that would otherwise be wasted. Now people can bake them into "bunny biscuits" for their animals.

Debbie started raising rabbits in 1983 with a mixed breed pair just for fun. She quickly discovered rabbit shows and started producing purebred Minilops. Holland Lops followed in 1985, when they were still hard to come by in New England. Over the years she has raised Blue-Eyed Netherland Dwarfs and Lionheads, but Hollands stole her heart and remain her favorites to this day. Her rabbitry, called Blackberry Farm, has garnered top placings at American Rabbit Breeders Association national convention shows, winning Best Breed Fur the first two times she entered. She has placed in the Holland Lop national shows as well. Nationally, she has served as Holland Lop

Rabbit Specialty Club director, treasurer, and president. In 2005, Debbie became the newest of only six inductees of the Holland Lop Club Hall of Fame in the organization's then 25-year history.

Debbie's first rabbitry was four outdoor hutches, enclosed on three sides. Her next was 15 hanging hutches in a building with half-walls that could be removed for the summer. Another effort resulted in 21 hutches inside a horse barn, but it was as cold as a barn (and that's pretty darned cold where she lives), so Debbie longed for a heated rabbitry.

For her current rabbitry, she had a concrete slab put down and a 24 × 32-foot building put up (see photo on page 81). It features nine windows, a front-entry door, and a 4-foot-wide door in the back that lets her wheelbarrow pass through. The building is insulated and has louver vents, as well as fans, in the front and back. She explained that the front fan pulls in air and the back fan draws it out.

A central oil furnace keeps the temperature above 47°F in winter, so Debbie does not have to deal with freezing water (or frozen fingers). That's a big deal in northern New England, where there are six months of winter and six months of poor sledding. Four vents send warm air below the hutches. The ceiling, 8 feet high, supports 16 inches of insulation for an R-value over 50. The walls are insulated to R-11, and all that insulation keeps the heating bills down and the summer heat out.

The oil barrel is in the right front corner. Above the barrel are two shelves of nest boxes marked with each doe's name. Originally, lengths of poly-ethylene plastic provided urine spray protection behind the two rows of back-to-back, two-high stackable hutches placed on 14"-high "tables," but most of them have since been replaced by Plexiglas. The tables minimize the amount of bending down required to care for the lower-tiered animals. Plastic protects the tables from any urine that doesn't make it into the dropping pans. The plastic gets washed but has to be replaced periodically.

A water pipe runs along the ceiling trim to the back, where a frostproof faucet is used to rinse off pans outdoors. Debbie keeps a hose cart by the back door and wheels it outside as needed. Dirty pans are scraped into a wheelbarrow and then hosed off outside and air-dried.

There is a stainless steel sink with running hot and cold water, a rug for grooming, and a supply cabinet. Shelving stores extra food and water crocks, tattoo supplies, and cups and bottles for the carriers she takes to shows. Feed is stored in an old chest-type freezer she received for free because the motor was dead. It keeps the feed dry, rodent-free, and cool. She stores most of her hay in the horse barn. A cart holds a feed bucket, a watering can, another bucket into which she dumps water from crocks, a small wastebasket, cleaning tools, and a box of hay.

"The back corner," she explained, "has an enclosed shavings bin [for her droppings pans] and a pop-out opening that allows us to shovel the shavings into the bin from the outside. You can drive the loaded pickup truck right up to the pop-out and stand in the truck bed and shovel shavings right into the bin. Two rows of fluorescent lights were centered over the two cage rows. Another strip was over the front counter area. One row went down the center."

She has since replaced the fluorescent lights with energy-saving CFL bulbs, which she leaves on during the day. She painted the walls a light blue to lighten the inside and stained the window and ceiling trim. An indoor/outdoor thermometer with a digital readout not only shows the temperatures but records the highs and lows indoors and out. A wireless intercom system to the house keeps Debbie and her husband, Gary, in touch.

"Although stacked cages are labor-intensive, I prefer them. We have 60 holes that are seldom completely full. It takes the two of us less than two hours a week to clean and sweep up. The building has worked out well for us. It is nice to go out to the rabbitry on a cold winter day and work in a sweatshirt, and it is also comfortable in the summer. I know I spend more time with my rabbits now, and I know my bunnies are more comfortable, too."

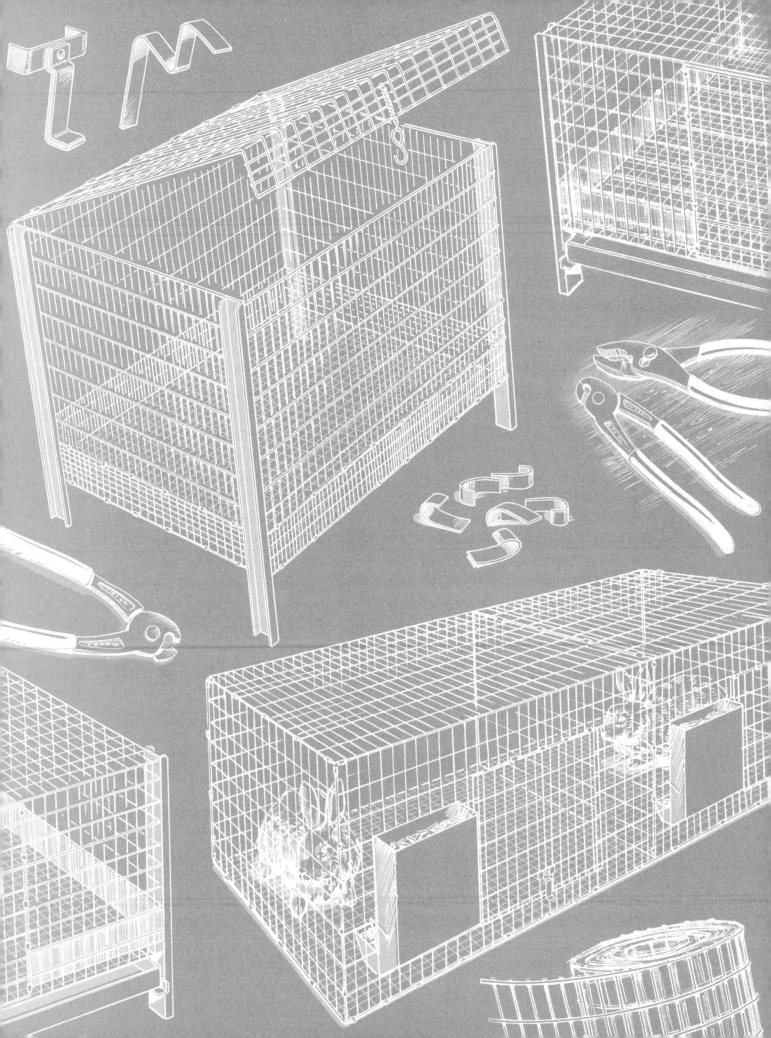

CHAPTER TWO

Building All-Wire Hutches

The welded wire hutch is basically a box made of 1" × 2" mesh for the four sides and top, with ½" × 1" mesh for the floor. Thanks to the mesh's rigidity, no other material is required for support, and the hutch can hold all the rabbits you can put inside. If you use it outdoors or in multiple tiers, it's best to locate the door in the front. If you use it indoors and on a single level, you can put the door on the front or on the top. Indoors, the hutch needs to be affixed to legs or suspended from rafters or joists. Outdoors, it needs legs and protection from the elements, with a roof and some sort of siding.

Making your own hutch is easy, and it is rewarding to make something that fits your needs and space exactly. You can buy the exact amount of mesh you want, but the cheapest option is to purchase 100-foot rolls. If you only want to make one or a few hutches, it isn't very difficult to find uses for the extra wire. For example, you could build a nonstandard-size hutch, use floor wire for cage fronts to help prevent small children from sticking their fingers inside, make a carrying cage, and even build a fence or trellis for your garden.

If you need only one or a few hutches and money is your primary concern, it's probably cheaper to buy a prefabricated cage from your local feed dealer or from a mail-order or Internet supplier. This is because unless you buy 100-foot rolls of mesh, the cost per foot is often more than the prefab hutch. The prefabs ship flat, are partially assembled, and can be completed in 15 minutes or less. There is no measuring or cutting, and no waste. They come complete with all the J-clips you need, plus the door latches and hangers. While prefab hutches come in standard sizes and are front-openers, some suppliers, such as Bass Equipment Company, will build custom sizes and even top-openers for little or no extra cost. Bass also supplies complete doors, ready to hang with the door lock in place, for use with custom-built hutches. KW Cages also offers custom cages, wire, and parts.

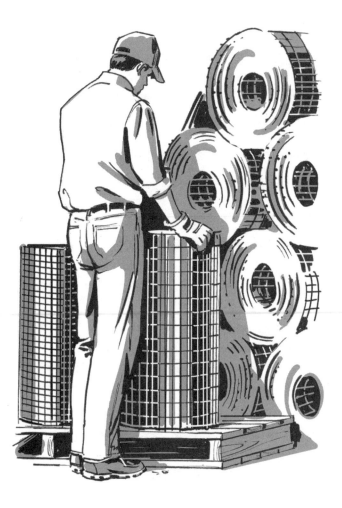

The most economical way of building hutches is to purchase 100-foot rolls of wire mesh, like those at this factory warehouse.

Choosing Welded Wire

There are a few different types of mesh to choose from: galvanized before welding, galvanized after welding, aluminum, and vinyl-coated. Galvanized before welding means the individual strands of steel wire are galvanized and then welded into mesh. Galvanized after welding means the strands are galvanized both before and after welding, which covers the weld joints and adds more galvanizing overall (it's also more expensive). Aluminized welded wire is steel wire that's coated with aluminum, which can make it last longer. Vinyl-coated mesh is the most expensive.

Reportedly, the *galvanized after* welding product lasts longer than the *galvanized before*, because welding the joints can eliminate some of the galvanizing there. Galvanizing "gives up," or oxidizes, over time, dissipating into thin air, so the more you have, the longer it lasts. I have used both types for 40 years or more, and both types are still in service. I've also used some of the aluminized mesh and don't see any difference. Overall, I don't think longevity is as much of an issue for the hutch as it is for the person who owns it. My hutches likely will last longer than I will.

I wouldn't use vinyl-coated wire to make a hutch. The rabbits will chew it, and the coating reduces the size of the openings to the extent that not all the droppings will fall through the floor. You might use it for a carrying cage because it's smoother and more comfortable to handle than the other types of wire. Plus, if a rabbit is inside

✳ ASSEMBLING A PREFABRICATED WIRE HUTCH ✳

This diagram shows how simply a prefab wire hutch is put together with J-clips. Simply secure the J-clips every 4 inches around the hutch.

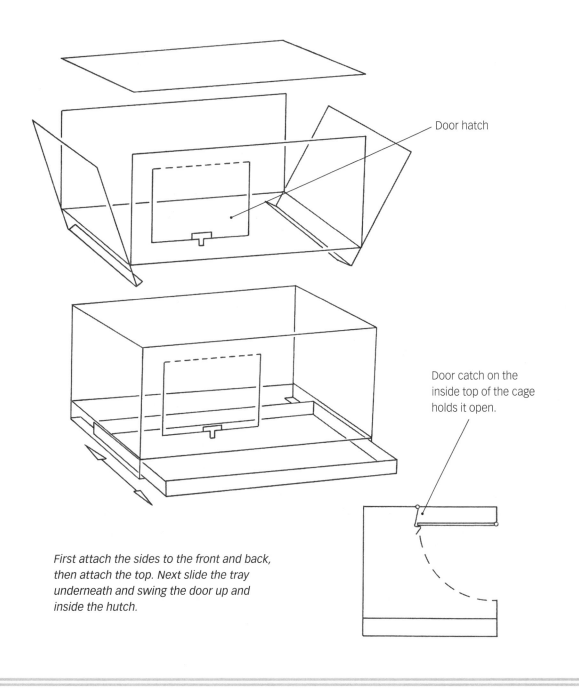

Door hatch

Door catch on the inside top of the cage holds it open.

First attach the sides to the front and back, then attach the top. Next slide the tray underneath and swing the door up and inside the hutch.

The Right Fastener

You can use J-clips and J-clip pliers or hog rings and hog-ring pliers to build your wire hutches. J-clips are available from farm supply stores and mail-order rabbitry supply companies, and hog rings are more widely available from farm as well as auto supply stores (they're also used to attach auto seat covers). C-rings are very similar to hog rings and are available in rabbitry supply catalogs.

When attaching urine guards, it is best to use hog rings (or C-rings). These rings are thicker than J-clips and thus more resistant to rust, which can be a problem when urine hits them. For this reason, you may also want to use hog rings to attach floors.

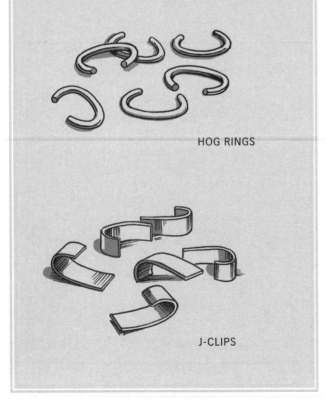

HOG RINGS

J-CLIPS

for only a few hours, such as for a show, it doesn't have much time to gnaw on the coating.

Be sure to use 14-gauge 1" × 2" mesh when building the sides and top. Anything lighter won't have the rigidity needed to hold up the hutch without supporting members. The floor wire can be either 14- or 16-gauge. The heavier 14-gauge wire is more rigid, while the 16-gauge is smoother and cheaper. Both will last a long time. For large rabbits, 14–gauge might be better for the floor.

Sizing the Hutch

Overall dimensions of hutches are up to you. You can make them in any size up to the maximum width of the wire, and even splice it with J-clips or hog rings to make them wider. There are several things to keep in mind, however.

You should provide nearly a square foot of floor space per pound of adult doe that will be producing a litter. A 2½ × 3-foot hutch will accommodate a medium-size doe — 9 to 12 pounds at maturity — and her litter to weaning time. A cage for a buck usually is about half the size of a cage for a doe and litter. You might build a 24" × 30" hutch for the smaller breeds or for a single pet or breeding buck. An 18" × 24" or 18" × 30" hutch will house a single pet dwarf very nicely. If you have one of the giant breeds, a hutch that is 30" × 60" is adequate. Another thing to consider: If you will be raising litters but plan to wean them at four to six weeks instead of the usual eight, your hutches can be smaller, but you will need additional hutches to grow them out.

The hutch depth should be no more than 30" from front to back; otherwise, you may not be able to reach rabbits in the rear. The height should be at least 14" for very small breeds and 16" or 18" for larger rabbits. For the giant breeds, some people make their hutches 24" high.

Welded wire mesh comes in widths of 12" to 72", in 2" increments. Keep these sizes in mind when planning your hutches to prevent unnecessary cutting and waste. Also, be sure to consider the space the hutch will occupy, whether indoors or out.

Planning the Door

The position of the door may be very important, depending on what you have inside and around the hutch. For instance, I locate the doors to my hutches to one side of the front to leave space for a self-feeder that will protrude through the wire. If you will use a water bottle, you'll need to provide room for that, too.

The door should be large enough to accommodate a nest box if the hutch is used for a doe and litter, and you should be able to reach inside to the far corners. If you use an all-wire nest box (my favorite kind, which is covered in chapter 3) of small or medium size, an opening 12" wide and 11" high is sufficient. Keep in mind that a hole that's too large can weaken the structural integrity of the hutch. For an 11" × 12" opening, I recommend a door that

is 12" × 14", so it can overlap the opening along the sides and bottom.

The door should swing up and in, rather than to the side and out. This way the door takes up no space in front of the hutch, and it won't obstruct an aisle or catch a sleeve when you reach inside. With just your thumb and a finger, you can swing the latch aside and push the door in, all in one motion, which can be handy if your other hand holds a watering can or a feed scoop. The best thing, though, is that if you forget to latch the door, it will not open if the rabbit pushes against it. Because the door overlaps the sides and bottom of the opening by 1", even a rambunctious rabbit won't be able to push it open under any circumstances. I can assure you that is not the case with doors that swing out.

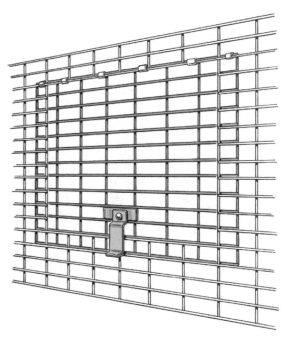

A hutch door opens in.

What about Plastic Flooring?

While I have been a proponent of all-wire floors ever since welded wire appeared in the 1940s, I have since become intrigued by the possibility of using perforated plastic flooring in conjunction with wire hutch sides and tops. What made me really think about it was the arrival of some very fancy Rex rabbits, a present from a West Coast breeder to whom I had sent some of my Tans. The sire and dam of these Rex rabbits, which came from imported English stock, had won Best in Show time after time, and I had high hopes for them. Rex rabbits have very short guard hairs, no longer than their underfur, including on the bottom of their feet. Unlike other, heavily padded breeds, Rex can develop sore hocks if raised on wire. I had so much trouble with them that I finally quit raising them, because I wasn't willing to put a solid, bedded floor under them.

Since then I have been looking for the perfect perforated plastic. An Internet search led me to FarmTek (see Resources). It offers a line of products for greenhouse operators and other growers that includes "PolyMax Poultry/Kennel Flooring," made of nonporous polypropylene. It comes in 24" × 48" black panels with ⅞" square openings, which are ideal for medium and large breeds. The panels are ½" thick and can be cut easily to fit your wire hutch.

Equipment suppliers also offer inexpensive plastic resting boards that you can place in a hutch to provide relief from the wire floor for rabbits, such as Rex, that might need it. These are made of a washable plastic with a smooth surface that is slotted to allow droppings to go through. You can remove the resting boards whenever it's time to burn hair off the floor of a wire hutch.

Here's how Polymax flooring can work with a hutch whose sides and top are made of wire.

PLAN: Easy Wire Hutch

➡ *Building a wire hutch is easy. You can't go wrong if you measure correctly because you're working with rectangular grids and everything comes out square and true. It might take an hour to build your first one, but soon after that you'll probably do it in half the time. There are several ways to go about building hutches, depending on how many you want. The following is one way to build a hutch that measures 36" wide × 30" deep × 16" tall.*

MATERIALS:

- 10 feet of 1" × 2" welded wire mesh, 14-gauge, 36" wide
- 3 feet of ½" × 1" welded wire mesh, 14- or 16-gauge, 30" wide
- 64 J-clips or hog rings
- Plastic clip-on strips or binder strips (optional)
- One door latch
- One door hanger

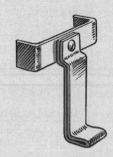

LATCH FOR DOOR THAT
SWINGS UP AND IN

HANGER THAT GOES ON
HUTCH TOP TO HOLD DOOR
UP WHEN YOU PUSH IT IN

TOOLS:

- Work gloves
- Tape measure
- Heavy-duty wire cutters (preferably flush-cut type)
- One 36" or 48" piece 2×4 lumber
- Hammer
- J-clip pliers or hog ring pliers
- Slip-joint pliers

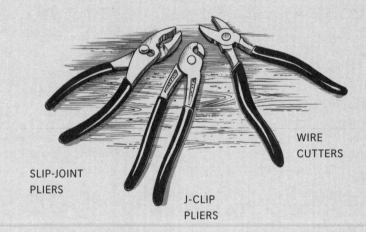

SLIP-JOINT
PLIERS

J-CLIP
PLIERS

WIRE
CUTTERS

A Good Building Environment

Anywhere with a concrete surface — such as a basement or garage floor, or a paved driveway — will make a good place to build an all-wire hutch. When working with welded wire, you might scratch a hardwood floor. A solid floor or pavement is the best place to bend the wire, too.

Cutting and Shaping the Wire

1. Cut the 14-gauge wire mesh into a piece that is 36" wide × 62" long, using wire cutters. As you cut, notice that a slight flick of the wrist down and away from the welds will snap off the wire cleanly with little effort.

Wire-cutting pliers that cut flush, leaving no sharp stubs, are the best kind for making hutches. Rabbitry supply houses sell them (see Resources).

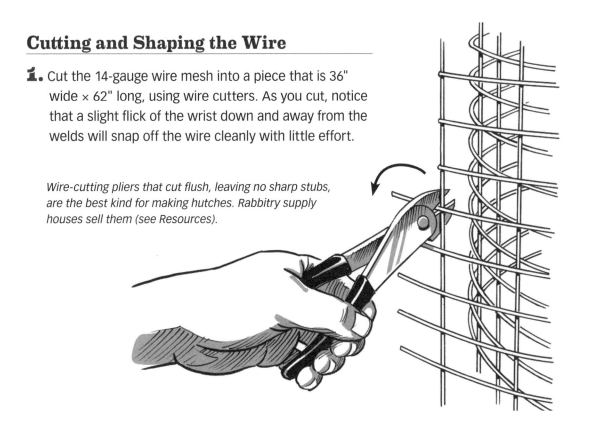

NOTE ON WIRE OPTIONS: Sometimes 1" × 2" wire is available as 2" × 1" wire, which simply means the 2" dimension is horizontal. Both types work, but keep in mind that every time you cut wire you lose 1" or 2" from the roll, depending on the type. You may also use 1" × 1" or ½" × 1" wire for the hutch sides and top (in addition to the floor), but these are more expensive than 1" × 2".

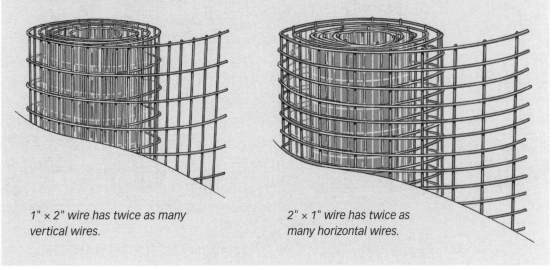

1" × 2" wire has twice as many vertical wires.

2" × 1" wire has twice as many horizontal wires.

Cutting and Shaping the Wire, *continued*

2. Flatten the mesh. Lay the piece on the floor (a garage or basement floor works well) so the curl from being rolled is convex (humps up). Stand on one end and gently bend the other end toward you, taking care not to kink it. You simply want to reverse the curve with enough tension to take the bend out of the piece.

If the roll comes with the wires spaced at 1" running on top of the wires spaced at 2", as viewed before unrolling (most do), then turn the piece over; this is important so that in the following steps you don't bend against the welded joints, but with them. If your wire is rolled the opposite way, be careful not to bend it against the welds because you could crack or break them.

3. With the wire on the floor, measure in 16" from one 36"-wide end and lay your 2×4 board across it at that point. Stand on the 2×4 and gently pull the 16" section toward you. Hold it vertically. Reach down with your hammer and gently strike each strand of wire against the 2×4 to fix it at a 90-degree angle.

FLATTENING THE MESH

FIXING WIRE AT A 90-DEGREE ANGLE

4. Measure 16" from the other end and repeat the process of bending and hammering. You have just formed the front, top, and back of the hutch. Set the wire aside.

5. Cut another 30" × 36" piece of wire mesh off the roll and flatten it, as before.

6. Measure in 16" from each 30" side of the piece and make the cuts to yield two 16" × 30" pieces; these are the sides of the hutch. The remaining piece is a 2" strip (plus stubs) from the center of the original piece. You will use this later, so set it aside.

7. Cut the floor: If your roll of ½" × 1" wire is 36" wide, cut a length of 30". If the roll is 30" wide, cut a length of 36". Do not flatten this piece of wire, which should hump up with the ½"-spaced wire on top, unless it has an extreme curl (this happens as you get close to the center of the roll). The idea is to have the ½"-spaced wires be uppermost, to provide a smooth surface for the rabbits and to keep an upward "spring" so the floor won't sag from the weight of the occupants.

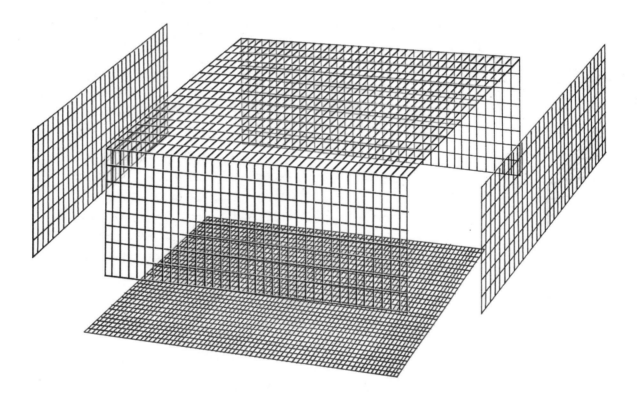

This exploded view shows how the wire sections will be clipped together to form the wire hutch. The door opening has not yet been cut.

Assembling the Hutch

1. Using J-clips or hog rings (and J-clip or hog ring pliers), fasten the front, top, and back section to the side pieces. Place a clip or ring every 4", starting with the corners. Make sure the vertical 1" wires are on the inside of the hutch. That way the horizontal wires are on the outside where they will make neat, tight corners when fastened to the front-top-back section. After a little practice you may find that you need to give the J-clips or rings a second squeeze with the pliers for a tight grip.

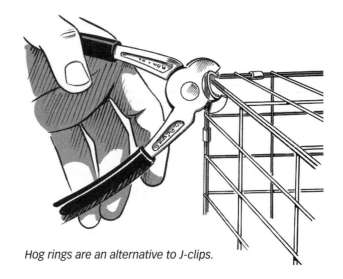

Hog rings are an alternative to J-clips.

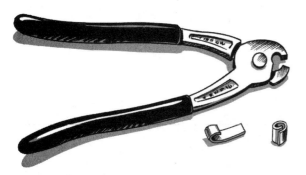

J-clips before and after clinching with J-clip pliers

2. Turn the hutch on its top and lay the floor wire over the bottom edges of the assembly so the convex curve and the ½" wires face down (toward the top of the hutch).

3. Clip or ring the floor wire to the sides, front, and back, starting in one corner and using a clip every 4". If the fit is too tight in a corner (this can happen if the 1" × 2" wire is made by a different manufacturer than the ½" × 1" wire), notch out the ½" corners of the floor wire by cutting out a ½" × 1" section with your wire cutters. You now have a sturdy and rigid box from what seemed flimsy at the beginning.

Cutting the Door

1. Cut the door opening by standing the hutch on its back with the front up and the floor closest to you. Beginning 4" from the floor, cut 12" across, up 11" on each side, and across the top. You can use this extra piece of wire to make a handy hay rack (see page 44).

2. Eliminate the sharp ends of the cut wire by leaving ½" stubs when you cut and bending them over with slip-joint pliers to create a smooth edge, or you can cut them off flush and cover the sharp edges with plastic clip-on strips (available from rabbitry supply companies). You can also use plastic binder strips made for binding school or business reports. Keep in mind that if you use a propane torch to burn off the hair on your hutch, you will need to remove the plastic protectors each time.

3. From the 36"-wide 1" × 2" wire: cut 1 piece at 12" × 14", making sure all the cuts are flush.

Attaching the Latch and Door

1. Position the latch 2" up from the bottom of the center of the door, fitting it over the wires that are 2" apart. Lay the door on a bench or the floor and flatten the latch over the wire with a hammer.

2. Fit the door inside the door opening and clip it to the hutch with J-clips or hog rings. Use a clip or ring on each end, and space three or four clips across the middle. Don't squeeze the clips or rings too tightly, so that the door is able to swing freely. The door should overlap the sides and bottom of the opening by 1", and the latch should swing easily (give it a drop of oil if it doesn't).

3. Attach a door hanger to hold up the door when it's pushed in. Swing the door up to the top of the cage and position the hanger 5" over from the left edge of the door. Secure the hanger by squeezing it with slip-joint pliers. The

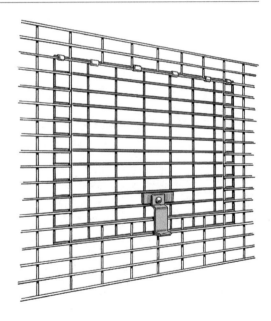

The door is positioned on the inside of the opening so it can swing up and in.

hanger will catch the door when it's pushed in all the way. To close the door, give it a little push; the hanger will swing loose and the door will shut.

Making a Baby Saver

If your hutch is for a doe and litter, it's important to prevent baby rabbits from falling out. Before they are old enough for their eyes to open, babies may hang onto the doe when she hops out of the nest box. They crawl blindly along the floor and are small enough to fit right through the 1" × 2" wire, falling to the ground or floor below. This baby saver feature keeps the babies safely inside the hutch until you can return them to the nest box.

1. Cut 2"-wide strips of 1" × 2" wire, one for each side of the hutch.

2. Position the strips along the bottom edges of the hutch, starting ½" above the floor. This reduces the size of the openings from 1" × 2" to ½" × 1". Attach the strips with J-clips or rings.

Stagger the 1" × 2" wire over the hutch sides to prevent the babies from falling through.

Making Urine Guards

Depending upon where you will locate your hutches, you may opt for urine guards instead of baby saver wire. Urine guards are strips of galvanized metal attached to the walls along the perimeter of the floor. You can use hog rings (or C-rings) to attach them to either the inside or outside of the hutch. They perform double duty, deflecting and diverting errant urine through the hutch floor rather than through the sides and to the ground below while also acting as a baby saver (eliminating the need for baby saver wire). If you are concerned about urine outside of the hutch, consider urine guards for *all* of your hutches, not just the ones for does and litters.

You can find urine guards that match the standard-size hutches sold by rabbit supply companies. One type is made for the insides of hutches, which I prefer, and the other is made for the outside. The inside type deflects urine down into pans, if you use them. The outside type keeps urine off the aisles but not necessarily in the pans. If you don't use pans, either type will work for you.

While manufactured urine guards are inexpensive, you can easily make them yourself. Buy a roll of aluminum flashing, preferably, or a sheet of galvanized iron from a hardware store or home center. Using tin snips, cut a strip of the flashing to the desired length (the combined length of all four sides of the hutch), then

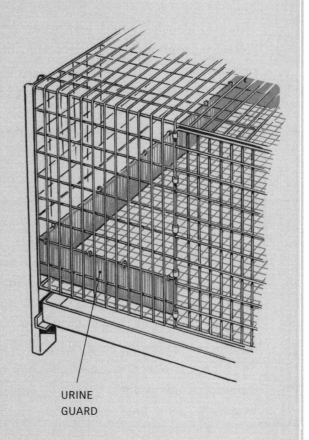

URINE GUARD

bend the strip to fit around the corners. If using a sheet of galvanized iron, cut a strip about the same width. You can even use roofing drip edge successfully. Secure the strip of flashing with sharp-pointed hog rings every 6" or so (the rings simply punch through the flashing). If you don't use flashing, you can punch some holes in the metal strip with a hammer and nail, then put the rings through it, or fasten the strip to the hutch with twists of wire.

Multiple Hutches

If you're planning a rabbitry of 10 or more hutches, it's a good idea to buy one roll of wire for the sides, fronts, and backs; one roll for the tops; one roll for the floors; and one roll for the doors — each at a specific width. You can avoid a lot of cutting by choosing widths that match or are close to the final size you need for each part of the hutches.

To form the four sides from one roll, bend a long strip of wire to three 90-degree angles to form a box, then clip the loose ends together to create the fourth corner. For dwarfs, you might use a 14"-wide roll for the sides; for small breeds, 16"; and for larger rabbits, 18". The size of the floor and top depend on the size of your rabbits. As mentioned earlier, the basic rule is to have about 1 square foot of space for each pound the mature doe weighs; hutches occupied by a buck should be about half that size.

To build hutches with multiple compartments, cut the side wire for the entire length of the hutch perimeter, and install partitions where you want them. Clip or ring the partitions to the front and back before installing the floor and top, then clip the partitions to the floor and top. Make the partitions from the ½" × 1" floor wire to eliminate the possibility of fighting or fur chewing between compartments. Alternatively, you can install double partitions of 1" × 2" wire spaced 1" apart so that rabbits can't reach each other with their teeth or toenails. A third option is to use galvanized iron sheeting for solid partitions. You can make these yourself by cutting 24-gauge galvanized iron with tin snips and punching holes for rings, or you can order some that are already cut to size and punched from a rabbitry supply house. Some hardware stores and home

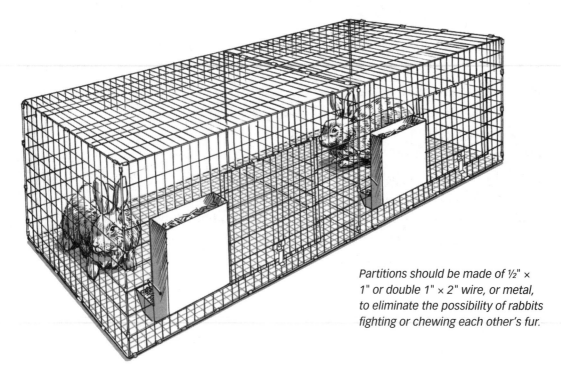

Partitions should be made of ½" × 1" or double 1" × 2" wire, or metal, to eliminate the possibility of rabbits fighting or chewing each other's fur.

centers carry perforated aluminum panels that you can cut to fit your hutch.

It's also possible to buy side wire with a built-in baby saver feature, where additional strands of wire have been welded to the bottom 4" of the roll. If you can't find it, you may want to buy 36"-wide floor wire for hutches that are 30" × 30", then bend 3" up on all sides. Clip these "flaps" to the hutch sides to create integrated baby savers. Alternatively, you can cut strips 3" or 4" wide and clip or ring them around the perimeter of the hutch, as described in Making a Baby Saver (page 28). Making your own baby saver is most efficient when building 10 or more hutches.

Drawbacks of Multiple-Compartment Hutches

Hutches with two or more compartments will not necessarily save you time and money over using single hutches. When ordering prefab multiple-compartment hutches, you may pay more for "oversize" shipping, particularly with hutches for large rabbits. Those for dwarfs or other small breeds may not cost you more. Also, multiple-compartment hutches often take just as much time to put together as the equivalent individual units.

Space and disposal are two other considerations. With multiple-compartment cages, you may need larger spaces if you decide to reconfigure your setup, which might be required to raise more rabbits. Individual single hutches are easier to fit into just about any size of row. Inevitably, the day will come when you need to get rid of the hutches. If you have single cages, you merely need to find a customer with a single rabbit. With multiple-compartment hutches, you'll likely have to find a breeder.

⟹ *If you plan to raise a small breed, one that weighs no more than about 6 pounds or so at maturity, you can build a smaller hutch. You need about 5 square feet for each doe, so you might make a single hutch 2 feet wide instead of 3, 30" deep, and 16 inches tall. Hutches of this size are also adequate for bucks or future breeding stock of the medium-size breeds. Here's what you need to build 10 hutches for small breeds.*

MATERIALS

- 100-foot roll of 1" × 2" welded wire mesh, 14-gauge, 14" or 16" wide
- 21 feet of ½" × 1" welded wire mesh, 14- or 16-gauge, 30" wide
- 22 feet of 1" × 2" welded wire mesh, 14-gauge, 30" wide
- 14 feet of 1" × 2" welded wire mesh, 14-gauge, 12" wide
- 5 pounds of J-clips or hog rings (see box on page 18)
- 10 door latches
- 10 door hangers

TOOLS

- Heavy-duty wire cutters
- J-clip pliers or hog ring pliers
- Slip-joint pliers
- Hammer
- 36" length of 2×4 lumber
- Tape measure

Note: This procedure is essentially the same as for building the Easy Wire Hutch (page 21). The main thing to remember is that you lose up to 2" when you cut 1 × 2" wire and up to 1" when you cut ½" × 1" wire.

1. Cut 10 lengths of 14"- or 16"-wide 1" × 2" wire mesh at 108".

2. As with building a larger hutch, you need to flatten the wire. Lay the piece on the floor (a garage or basement floor works well) so the curl from being rolled is convex (humps up). Stand on one end and gently bend the other end toward you, to reverse the curve without kinking it. If the roll comes with 1" wires on top of 2" wires, flip it over before proceeding. If your wire is rolled the opposite way, don't bend it against the welds because you could crack or break them.

3. Lay each piece of wire on the floor. Measure 24" from one end and position the 2×4 board across it at that point. Stand on the 2×4, gently pull the 24" section toward you, and hold it vertically. Reach down with your hammer and gently strike each strand of wire against the 2×4 to fix the mesh at a 90-degree angle.

4. Bend the wire again at 54" and 78". Using J-clips or hog rings, fasten together the loose ends of the wire to form a box. When fastening, place a clip or ring every

4", starting with the corners. Make sure the vertical wires are on the inside of the hutch, so they will make neat, tight corners when fastened. You may need to squeeze the J-clips or rings a second time for a tight grip.

5. Cut 10 lengths of 30"-wide ½ × 1 wire at 24" for the floor. Do not flatten these pieces of wire. They should hump up with the ½"-spaced wire on top, to provide a smooth surface for the rabbits and to keep an upward "spring" so the floor won't sag from the weight of the occupants. Secure each piece to one of the hutch boxes. If the fit is too tight in a corner, cut out a ½" × 1" section of floor wire with your wire cutters.

6. Cut 10 lengths of 30"-wide 1 × 2 wire at 24", flatten as above, and clip or ring each to a box to create the tops.

Cutting the Door

1. Using the 12" wire, cut 10 pieces at 12" × 14" for the doors. Make sure all of the cuts are flush.

2. Cut the door openings by standing the hutches on their back with the front up and the floor closest to you. Beginning 2" from the floor, cut up 11" on each side and 12" across the top, so that the door is large enough to accommodate the nest box. The extra piece of wire can be used to make a hay rack (see page 44). Be aware that, whether the door is on the left or right, you will have less space next to the door than you do in a larger hutch. That won't be a problem because smaller hutches mean either smaller or fewer rabbits, and you will use a narrower feeder than with the larger breeds.

3. To eliminate the sharp ends of the cut wire, leave ½" stubs and bend them over with slip-joint pliers. Alternatively, you can cut them off flush and cover the edges with plastic clip-on strips (available from rabbitry supply companies) or plastic binder strips made for binding school or business reports. Keep in mind that you will need to remove the plastic protectors each time you clean the cage with a propane torch.

Attaching the Latch and Door

1. Lay the door on a bench or the floor. Position the latch 2" up from the bottom of the center of the door, fitting it over the wires that are 2" apart, and flatten it with a hammer.

2. Fit the door inside the door opening and clip it to the hutch with J-clips or hog rings. Use a clip or ring on each end, and space three clips across the middle. Don't squeeze the clips or rings too tightly, so that the door is able to swing freely. The door should overlap the sides and bottom of the opening by 1", and the latch should swing easily (if not, give it a drop of oil).

3. Swing the door up to the top of the cage and position a door hanger 5" over from the left edge of the door. Secure the hanger by squeezing it with slip-joint pliers. The hanger will catch the door when it's pushed in all the way. To close the door, give it a little push; the hanger will swing loose and the door will shut. If desired, you may add the baby saver feature or urine guards.

➡️ *A top-opening hutch uses its entire top panel of wire for a door. These are appropriate only for single-tier setups, but they offer some nice conveniences. They're ideal for indoor use and can also be placed outdoors if you incorporate a waterproof roof over the wire top. Without the door taking up wall space in the front of the hutch, you can position hay and pellet feeding and watering equipment just about any place you like.*

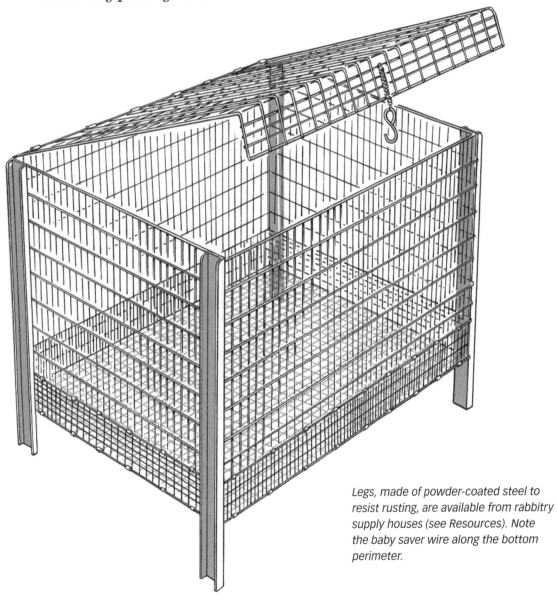

Legs, made of powder-coated steel to resist rusting, are available from rabbitry supply houses (see Resources). Note the baby saver wire along the bottom perimeter.

With a single-tier arrangement of top-opening hutches, you can set them at a lower height than with front-openers, making it easier to check the amount of feed in the feeders. Manure disposal is also simpler: It falls below to the floor, the ground, or whatever you use for collecting it. With the hutch at a handy height of about 2 feet above the floor or ground, you can easily lift the top to catch a rabbit inside. This also makes it easier and quicker for positioning a nest box, and you can see inside the box without moving it or disturbing its occupants. For these reasons, some of the biggest commercial rabbitries have long favored the top-opener.

When it comes to construction, top-openers require less wire because there's no door. They're also easier and faster to build and require fewer J-clips or rings, plus they don't need conventional latches and door hangers. This plan shows you how to build an all-wire top-opening hutch that measures 30" × 36" × 18".

Note: If this is your first time building hutches, see the Easy Wire Hutch (pages 21–27) for detailed instruction on the basic steps of creating an all-wire hutch.

MATERIALS

- 11 feet of 1" × 2" welded wire mesh, 14-gauge, 18" wide (100-foot roll makes the sides of 9 hutches)

- 2½ feet of 1" × 2" welded wire mesh, 14-gauge, 36" wide (100-foot roll makes the tops of 32 hutches)

- 3½ feet of ½" × 1" welded wire mesh, 14- or 16-gauge, 36" wide (100-foot roll makes 28 hutches with baby saver feature)

- J-clips or hog rings (see box on page 18)

TOOLS

- Tape measure
- Heavy-duty wire cutters
- 36" piece 2×4 lumber
- Hammer
- J-clip or hog ring pliers
- Slip-joint pliers

Quonset-Style Hutches

Some people have built Quonset-style hutches with rounded tops that include the door. They are usually built so that the required minimum height is in the center, with the top curving toward the floor at both front and rear. One variation is called a semi-Quonset, where the back wall is of the required height and curves down to the front.

Quonset hutches appeal to some because they require less wire, but it takes more time to cut the wire on the sides. Depending on what you have for feeders and watering equipment, the down-curving front can complicate installation. If I were you I wouldn't fool around with them.

1. Flatten the 18"-wide piece of wire. Lay the piece on the floor so the curl from being rolled is convex (humps up). Stand on one end and gently bend the other end toward you, taking care not to kink it. If the roll comes with the wires spaced at 1" running on top of the wires spaced at 2", as viewed before unrolling (most do), turn the piece over before proceeding, so that you don't bend against the welded joints. If your wire is rolled the opposite way, be careful not to bend it against the welds because you could crack or break them.

2. Bend the wire at 90-degree angles to form a box. Place the wire on the floor and lay your 2×4 board across the wire 36" in from one end. Stand on the 2×4 and gently pull the 36" section toward you. Hold it vertically. Reach down with your hammer and gently strike each strand of wire against the 2×4 to fix a 90-degree angle. Repeat this at 66" and 102". You can also use a work bench or rectangular post in place of the 2×4 on the ground.

3. Clip the loose ends of the piece together with J-clips placed every 4", starting with the corners, to complete the side assembly. You may need to squeeze the J-clips a second time for a tight grip. Make sure the vertical wires are on the inside of the hutch, so that your corners will come together nicely.

4. To make the floor, cut a piece of 36"-wide ½" × 1" wire at 42". Do not flatten this piece of wire. It should hump up with the ½"-spaced wire on top, to provide a smooth surface for the rabbits and to keep an upward "spring" so the floor won't sag from the weight of the occupants. To make integrated baby saver sides, make 3"-deep cuts at each corner, then fold up the "flaps," and clip or ring the floor to the hutch box (this step is similar to making a dropping pan from sheet metal).

5. Make the top by cutting 34" of 36"-wide 1" × 2" wire and flattening the piece as described in step 1. Using a hammer and 2×4, create a 90-degree bend 4" from the front edge of the top, to form a front "lip" or flange.

6. Clip or ring the top piece to the top rear edge of the box, keeping the fasteners a little loose so they work like hinges (I prefer rings here). In place of a latch to keep the top securely closed, you can use a spring and an S-hook or a dog-leash snap fastener.

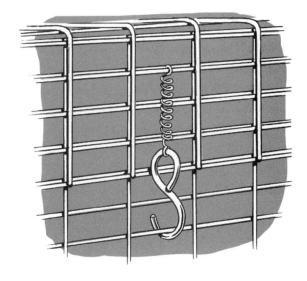

SPRING AND S-HOOK

A Man with a Plan *in Colorado Springs*

Myke McCune

LOCATION: Colorado Springs, Colorado

AVERAGE RABBIT POPULATION:
Six breeders, plus litters

GOAL: Home meat supply

BREED: Californian

MYKE McCUNE RAISED rabbits for awhile as a kid in the country. Then he moved to town, married, had children, and never thought about rabbits again until one day, on a trip to the library, he came across one of my rabbit books. It renewed his interest and answered some of his questions:

"With the cost of living going up and my paycheck staying the same, I asked myself, how could I cut back on the grocery bill? How could I have a successful garden with the sandy soil in my yard? That night, I read about the benefit of rabbit manure for fertilizer, as well as the low cost and nutritional value of meat rabbits. I decided to start raising them.

"I did a lot of research to determine what I would need to provide a proper home for my herd, which I decided would consist of Californians. I priced kit barns at the local hardware store and decided it was cheaper to build than to buy. It took quite a bit of explaining to my wife how spending money on a shed, cages, breeding rabbits, and feed was going to save us money, but she finally agreed. I was off to the lumberyard.

"With a little help from my daughters, Nancy (age 18, a college nursing student) and Faith (age 11), I had a shed built in three days. It is 8 feet wide and 12 feet long. Eventually I will have three sets of double-stacked cages along the 8-foot walls. I plan to have two bucks and four does. The remaining six cages will be for growing young rabbits to butchering age."

Myke made a 4-foot opening in the back wall and covered it with lattice to help ventilate the shed. The door is also made of lattice. He added plywood sliding doors that he could close when it's windy and on cold winter nights. The shed is situated to take advantage of the breeze, which usually blows from north to south. There is a 4-foot space between the west side of the shed and the neighbor's retaining wall and fence. He also built an abattoir from a 55-gallon drum and will use it outdoors, except in the winter, when he will move it into the garage.

> "With the rabbits for meat, the chickens for eggs and occasionally for meat, and both providing free manure to add to my sandy soil and garden beds, I expect our grocery bill to drop."

Myke is also looking at other sources of food and sees his rabbits as an important piece of his growing homestead. "I also built a chicken coop using the south wall of the rabbit shed. That allowed me to use the shed and coop for two sections of the chicken yard and to fence in an 8-foot-square area. The east side wall will get furring strips and a lattice covering, and I'm going to plant raspberries there. With the rabbits for meat, the chickens for eggs and occasionally for meat, and both providing free manure to add to my sandy soil and garden beds, I expect our grocery bill to drop."

He will build six garden beds, each 3 × 16 feet. Eventually he hopes to sell excess manure to local gardeners and to trade surplus meat rabbits, eggs, and garden produce to friends and neighbors for other food items.

"I have finished the trim and put up the fiberglass wall panels. After thawing water bottles many times last winter I decided to remove the fiberglass, add insulation, and replace the fiberglass. I am also planning a watering system. I will add plastic pipe to all the cages. The water tank will use a small pond pump to circulate the water through the pipe and back to the holding tank; the moving water won't freeze. I may add an immersion heater if I need it.

"I will probably go with fiberglass dropping boards instead of pans. I also need to fence in the chicken yard and put the final touches on the coop. The Californians I bought are purebred and pedigreed and come from a good line, so I also plan to sell show and breeding stock to help make more money," he says.

Myke also plans to tan the hides and make a rabbit-fur blanket, with an eye toward eventually selling the blankets or just the hides. Myke's plan to decrease his food costs, improve his garden and yard, and make some extra money — all centered on rabbits — is a plan that has worked for many others. But it has required a family commitment.

"My wife, Stephanie, a professional photographer who also works in a doctor's office, and my daughters have been supportive and help take care of the rabbits and chickens while I work nights."

During the night he is a supervisor at a company that fabricates metal components for the railroad industry — not the trains, he says, but just about anything that goes on them. With all that he plans to do at home, it looks like he will be very busy during the daytime, too.

CHAPTER THREE

Outfitting the Hutch

Before we discuss where and how you will place your hutch, let's talk about supplying it with feeding and watering equipment and, for the doe hutches, nest boxes. These three items are really all you need with the hutch.

As discussed in chapter 2, I suggest placing a door on one side of the hutch, not in the middle, and have it open inward. Putting it on one side leaves plenty of room for a self-feeder and either a water bottle or water valve. If you use a crock or water pan, there is room for that, too, without blocking the door opening. Furthermore, if you use certain types of automatic watering, a door that swings in will make it easier to install piping or tubing. If you place the door in the middle, and either the door is large or the hutch is small, you'd probably have to make the door swing out to the side because it would have to hold the self-feeder. The weight of the feeder and feed can cause the door to sag or even break off over time, and you run the risk of spilling feed when you open the door.

A door to one side also facilitates insertion and removal of nest boxes, which are best placed in a rear corner; you can just slide the nest straight in and out through the door.

Types of Feeders

The self-feeder, sometimes called a hopper or a J feeder due to its shape, attaches to the outside of the hutch and requires an opening in the wire to allow the trough to protrude inside. For small and medium-size breeds, it's a good idea to position the trough 4" to 6" above the floor. If it's lower than this, rabbits tend to get their front feet into it and scratch out the feed pellets, which drop through the floor and are wasted. Don't worry about the young rabbits; when they are old enough to eat pellets, they will stretch on their hind legs and be able to eat from the trough. Very young ones have been known to hop right in.

The self-feeder has a lot to recommend it. It takes up no floor space, it can't tip over, the trough protrudes only a couple of inches into the hutch, and it prevents all but the smallest babies from sitting in it and potentially fouling it with feces. A "lip" on the edge of the trough also helps keep the pellets from being scratched out and wasted.

Self-feeders come in various widths, from 3½" for single rabbits up to 11½" for does and litters of the larger breeds, with three sizes in between. You can choose either a solid or a screened bottom. I like the type with a screened bottom because the screen lets any pellet dust fall through, making the feeder self-cleaning and also preventing the rabbits from wasting feed by scratching out the feed lying under a coat of pellet dust. The screen also makes it easy to see if the rabbits have eaten all of the pellets, which helps you to determine how much to add. Bass Equipment Company (see Resources) has sold the screened-bottom self-feeder for about 50 years, and I haven't seen a better one yet.

Hay/Pellet Feeders

One type of self-feeder comes with a built-in hay compartment and is called a hay/pellet combo feeder. In the interest of full disclosure, this feeder was developed by a Dallas, Texas, rabbit breeder named Phil Angell, who modified a regular self-feeder. He sent me his homemade version, and I sent it to Bass Equipment Company. They liked it so much they decided to manufacture it, and it earned a prominent spot in their catalog. It's my favorite kind. It allows you to feed dry hay and its leaves, which get caught in the feed trough. The most nutritious and tastiest part of alfalfa and clover hays, for example, are the little leaves, but they often crumble and fall through the hutch, leaving only the stems. It's a good idea to cut hay into short lengths, about 6", because rabbits tend to pull out longer lengths, eat some and then let it go, thus wasting it through the hutch floor. If you raise rabbits with especially large heads, such as Flemish Giants or French Lops, a feeder with a wide "mouth" or trough accommodates them nicely.

The fact is, you don't really *need* to feed hay because rabbit pellets are about 50 percent forage products, but it does add fiber and gives the rabbits something to munch on. And it's true that you really don't have to use any kind of hay rack with the all-wire hutch. You can simply put some hay on top and let the rabbits stretch up and pull it through. But if you wish to feed hay in a container other than the hay/pellet combo feeder, you might use a hay rack.

Prefab galvanized metal hay racks attach to the front of the hutch. Before those were available, I used to fashion hay racks from cage wire. These were "boxes" with the same width and depth as the self-feeder trough. I attached them to the inside of the hutch right above the self-feeder, high enough so it didn't obstruct the trough. I cut out a section of the hutch wire to give access to the inside and stuffed the hay in there. Uneaten leaves would fall to the feeder trough. I also made some simple hay racks with small scraps of 1" × 2" wire, such as leftovers from making door openings.

Making your own hay rack is a simple project: Bend a 6" or 8" square scrap of 1" × 2" wire mesh 2" from the edge, to an

* FEEDERS *

You can't beat the J feeder, which is available in several types and sizes.

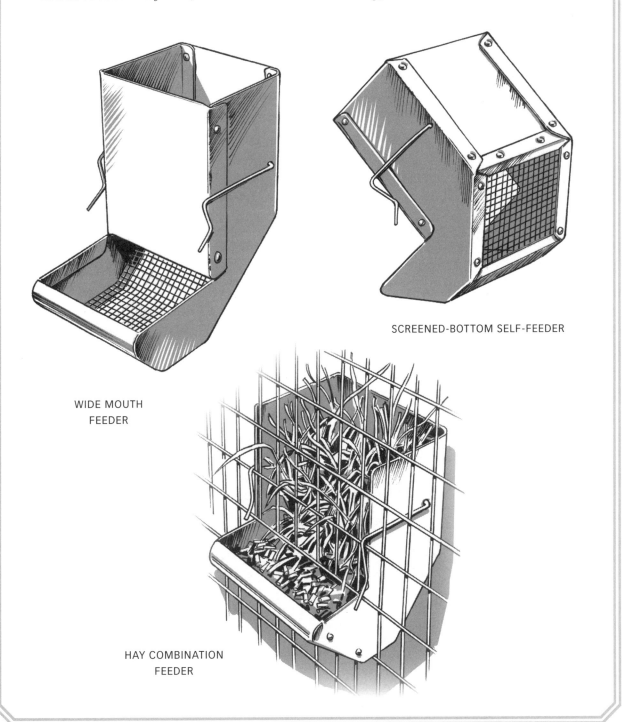

WIDE MOUTH
FEEDER

SCREENED-BOTTOM SELF-FEEDER

HAY COMBINATION
FEEDER

❋ HAY RACKS ❋

You can buy a manufactured hay rack, or easily turn leftover scraps of 1 × 2 wire into a hay rack.

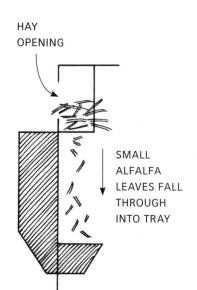

HAY OPENING

SMALL ALFALFA LEAVES FALL THROUGH INTO TRAY

Position the feeder below the hay rack (right) so that small pieces and leaves of hay fall into the feed trough below (above).

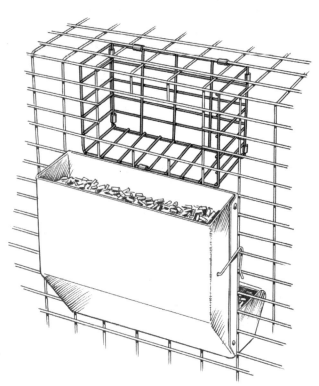

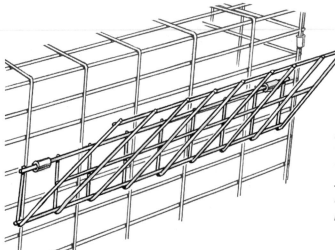

You can make a very simple hay rack by bending a scrap of welded wire and fastening it to the hutch with a couple of J-clips. Just place a handful of hay on the rack and the rabbit pulls it through the wire.

angle of about 30 degrees. Using a couple of J-clips or hog rings, fasten the 2" side to the front of the hutch wherever it is convenient. Fill the rack with a handful of hay; the rabbits will pull it through.

Creep Feeders

Another kind of feeder that I tried for a few years, and some people still use, is the creep feeder. The idea is that young rabbits just out of the nest box benefit from a special feed that only they can access. The creep feeder has a trough with small openings that keep the doe out. After using it with a great many litters, I decided that creep feeding held little if any benefit for my rabbits. They grew just as well on the regular feed, so I stopped using the feeders and quit buying the more expensive creep feed. I have since turned the creep feeders into bird feeders that pretty effectively let the little guys in and keep the crows out.

Buying and Using Self-Feeders

I can't say enough about the self-feeder. For years, rabbit raisers used crocks. But if you order crocks to be shipped to you, the cost is a lot more than with the metal self-feeder, and the crocks can break en route. Also, rabbits can defecate and urinate in the crock, introducing a health problem, wasting feed, and causing extra cleaning work.

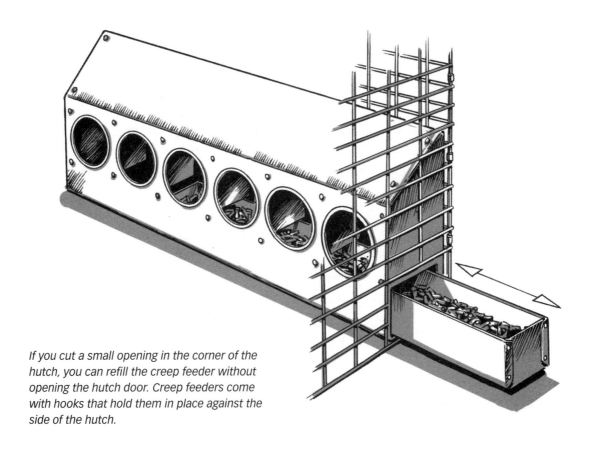

If you cut a small opening in the corner of the hutch, you can refill the creep feeder without opening the hutch door. Creep feeders come with hooks that hold them in place against the side of the hutch.

Hidden in Plain Sight *in the City*

Green Meadow Farms
R. L. Wilson

LOCATION: Omaha, Nebraska

AVERAGE RABBIT POPULATION: 200

GOAL: Meat rabbit business

BREEDS: New Zealand White, New Zealand Black

R. L. (DICK) WILSON calls his Nebraska rabbitry Green Meadow Farms and advertises "Wholesome, Humanely Raised, Chemical-Free Domestic Rabbits." He has been raising rabbits since he was a boy in 1949. A former game warden, he now owns an architectural millwork business.

I imagined Green Meadow Farm as an oasis amid Nebraska cornfields stretching flat to a brown and yellow vanishing point innumerable miles away. But Dick's rabbit farm is nowhere out there. It lies inside what passes for a cornhusker metropolis, the state's largest city, Omaha.

"I raise and sell top-quality commercial New Zealand Whites and also a few purebred New Zealand Blacks, some of which I give to 4-H kids," Wilson says. He sells the Whites for meat, both live and dressed. Some of those he butchers reach restaurants and stores. Among his other customers are heart patients, who value the low cholesterol, low fat, and high protein, and the city zoo, which uses the rabbits to feed some big cats. Snake breeders also buy some.

The rabbitry is a remodeled little barn (see photo on page 81). The new structure features lattice in front of solid sliding panels that open for ventilation. Glancing at the entrance arbor, you'd think it's a garden house. His raised-bed vegetable garden along one side convincingly supports the ruse.

"I'm in the city, and over the years the neighborhood grew up around me. When a local woman complained about the looks of the barn, I turned it into a garden tool house. I told her the neighborhood no longer needed to tolerate a barn. Instead, it is home to a 'hoe house.' What I really have now is a rabbitry hidden in plain sight in an urban neighborhood," Wilson explains.

Inside his rabbitry, which features plenty of lighting, a big fan, and no heater, you'll find two types of wire hutches housing up to 200 rabbits. On the left is a single tier with the tops sloped to the front. Doors open on top. "That makes it easy to reach any animal, and if the door is open the frisky young ones stay inside," he told me. Behind the hutches are rolls of rubberized roofing material. It hangs behind them to protect the walls from urine. In extremely cold weather he also rolls them forward over the hutches and down to the floor. That helps hold in the rabbits' body heat and prevents drafts. Above the hutches, near the ceiling, are handy electrical outlets so he can plug in nest box heating pads or additional lighting if needed.

Wilson prefers to use water bowls made of fiberglass because he says he can knock the ice right out of them with no fear of breakage, and he doesn't have to worry about tongues sticking to metal during the winter.

"The double stack on the right saves space, but dropping pans under the top tier are a big nuisance. They are difficult to handle even when only half-full and are hard to carry out," said Wilson. Conveniently, though, the droppings go just a few steps to the garden that helps keep the secret of the hidden city rabbit farm.

You will notice every day how convenient the self-feeder is because you don't have to open the cage to administer the pellets. You might use a small scoop to put in the pellets or, like I do, learn how many handfuls each rabbit requires. Self-feeders are made of galvanized steel, but even if they were made of sterling silver they would be worth the money in the long run over tin cans or crocks because of the amount of feed they prevent from being wasted (not to mention the convenience). They also give the rabbitry a professional appearance, which is especially important if you want to sell your animals.

Both local farm stores and rabbitry supply companies offer self-feeders. The feeders come with wires for easily attaching them to the hutch. You can also order stainless steel attachment springs for easy removal and replacement for periodic cleaning, but if you use the screen type you will hardly ever have to take them off. Another optional accessory is a cover designed to keep out rain or snow or even errant urine from a hutch above. However, you have to lift these up every time you put the feed in, then put them down again. There are better ways to protect the feeders, including providing better overhang protection for the entire hutch.

Watering Methods

Rabbits need access to water all of the time, and there are various ways to deliver it. Some are excellent, and some are not so good. Let's look at them one at a time.

TIN CAN. This could be a one-pound coffee can. The design is simple, but it has some serious drawbacks. If the can is left free-standing, rabbits love to tip it over, then they go thirsty until you refill it. You could punch a hole in the side of the can and wire it to the side of the hutch (I did that years ago), but you have to unhook the can to rinse it out, and the can protrudes halfway into the hutch. Over time it will rust, of course, and need to be replaced. Also, rabbits are bound to foul a can periodically, making it an open sewer.

CROCK. This won't tip over, but it will freeze and break unless the inside diameter at the bottom is smaller than the inside diameter at the top, so that ice will slide upward instead of push outward. Crocks aren't cheap, they break if dropped, and they rival the coffee can for open sewer honors.

A crock shaped like this won't break if the water freezes.

METAL AND HARD PLASTIC water vessel. This device attaches to the cage wire above the floor, where it is less likely to be fouled and where it doesn't take up floor space. You have to reach inside and take it out to rinse or wash it, but it is a big improvement over the can or the crock on the floor. This vessel can be especially useful during freezing weather because ice won't break it. You can smack the ice out of the vessels by knocking two of them together, or you can dislodge ice by slipping it into a bucket of warm water. Many rabbit keepers have two for each hutch and keep one inside where it's warm to replace the one that's frozen. You can also fill the vessel partway in the morning, let rabbits drink until the water freezes, and then fill it the rest of the way in the evening. Ice won't stop the containers from functioning completely, as it does with water bottles.

WATER BOTTLE. You can buy plastic water bottles with rubber seals and metal drinking tubes for just a few dollars. Because the bottle is enclosed, the water does not get contaminated. The bottles are affixed to the cage with wire brackets, and you

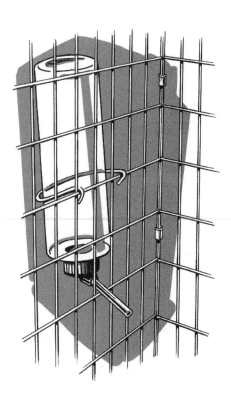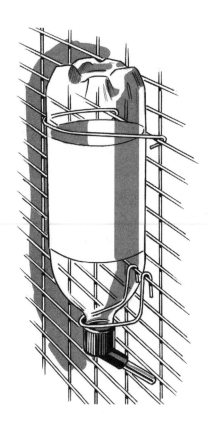

You can purchase a one-quart water bottle (at left) from rabbitry suppliers or from your local farm/feed store. Or, you can just buy a cap and tube and use it on a plastic soda bottle (at right).

unscrew a plastic top to refill them. You can also purchase top-filling bottles with better brackets that you can refill without taking off the cage. Bottle capacity ranges from 8 to 64 ounces.

Another type of water bottle can be made with a 2-liter soda bottle and a conversion kit that includes a tube and cap assembly and a special bracket and spring. One of the best conversion kits features a spring-activated drinking valve like the ones used in automatic watering systems (see page 50). You can also buy these valves individually and make your own: With a knife or drill, make a hole near the bottom of a 2-liter plastic bottle. Coat the threads or the barbed bottom of the valve with an epoxy cement and screw or push it into the hole. Use a loop of wire to hang the bottle onto the hutch. You can use more than one per hutch if necessary.

Another option is to insert one of the barbed valves into a length of black vinyl tubing and connect it to a bucket or large plastic bottle that sits on or hangs above the hutch. The connectors, available from rabbitry supply houses, can be attached to any plastic bucket or tank by simply drilling a hole. They include a strainer, a shut-off valve, and a barbed outlet. If you use a large vessel, it can serve several hutches via tubing, fittings, and valves, none of which cost very much.

As mentioned, below-freezing temperatures can knock a water bottle out of business. You can get around that with a bucket and tubing, an electric bucket heater from

Salt This Advice Away. Far Away.

Salt spools are commonly available in feed and pet stores, but your rabbits don't need them and your hutches don't need them. Check the ingredients tag on your feed bag. Salt is in there already, and rabbits need no more salt than that. Also, salt spools drip with condensation and will rust your hutch floor just like road salt rusts the floors of Volkswagen Rabbits.

Salt spools were developed many years ago, in the days before complete rabbit rations and wire floors. If, despite this advice, you become absolutely convinced that your rabbits need additional salt, then shake some onto their pellets. That way you will rust out only the feeder, which is a lot cheaper and easier to replace than the hutch floor.

a farm store or a tropical fish aquarium, or an immersion heater from a pet store. Of course, a large vessel will need less refilling, but you don't want it to be so large that the heater won't keep the water from freezing. A good thing about the use of plastic bottles, tubing, and valves is that freezing temperatures won't break them.

SEMIAUTOMATIC AND AUTOMATIC WATERING. The best way to deliver water is with a pipe or tubing watering system. Nothing beats a constant supply of fresh water.

An automatic system sends a continuous supply of water to a tank via either a pipe plumbed to your household water or a hose from an outdoor spigot (hose faucet). In a semiautomatic system, you fill a larger tank by hand with a garden hose, a watering can, or a bucket. If you live in a cold-weather climate and use an automatic system with a hose or pipe running along the ground, you might want to switch to a semiautomatic system during freezing weather, simply by filling your tank by hand. That's how my first system worked: It was automatic most of the year but semiautomatic during freezing temperatures. I used a 5-gallon tank, and that worked well for 36 hutches if filled twice a day.

An additional benefit of either system is the ability to medicate water if necessary. You simply measure the amount of a liquid medication or a soluble powder product, such as Terramycin, for the capacity of your tank when full. Then, shut off the incoming supply until the tank is empty. Alternatively, you can buy a proportioner that lets you medicate the water without shutting off the incoming supply, but this is expensive and you really can get by without one.

Installing a Watering System

I installed my first watering system in 1970. I've since put in another system and have consulted on a great many more. There really is very little mystery to this. You need a source of water and a holding tank or bucket. If you are using an automatic system where the water is supplied continuously to the tank, the tank doesn't need to be very big because all it really does is break the water pressure. The water flows by gravity to drinking valves in each hutch. The water pressure can be changed simply by turning the valves; that way you can ensure the rabbits get just enough water to drink, but there isn't so much pressure that the water could squirt them in the eye when they bite on the spout.

In an automatic system, the tank has a float valve like the one in your toilet tank. As the rabbits drink and the water level recedes in the tank, the float valve opens and admits more water from the source. This provides a continual supply of fresh water. You can also use a pressure-reducing valve instead of a tank but only if you're not concerned about freezing temperatures, because ice could disable or even break the valve.

If you have a small rabbitry with only a couple of dozen hutches, watering by hand during the coldest time of the year is not a lot of work. To make it even easier, you might utilize a portable electric space heater when it gets really cold. My own barn has a small greenhouse attached to the south side, and while it is unheated and uninsulated, the warmth from the greenhouse (when the sun is shining) mitigates the cold. If you have a large operation in the northern part of the country, you will want to have a building that is insulated and perhaps heated. In chapter 5 we will discuss the construction, insulation, heating and, most important, ventilation of buildings. After all, keeping rabbits cool is a lot more important than keeping water warm.

Flexible Tubing System

After you decide whether you'll fill your tank automatically or manually, you will need to determine how to get the water from the tank to the hutches. Let's start with the system I like best: the flexible tubing system.

As you can see from the illustration on page 52, the tubing isn't the only thing that's flexible. The main advantage of this system is that you can expand and reconfigure it quickly, easily, and cheaply as your rabbitry grows. It costs less than alternative watering systems, it installs much easier, and the only tools you need are a pair of household scissors and a tape measure. You can place the drinking valves on the front of the hutch, no matter what kind of door you have. The only disadvantage to the system is that it limits your options for keeping it running during freezing temperatures, but in my opinion it's still the best way to go.

To install a flexible tubing system, first position your tank, bucket, or whatever vessel will hold the water above the highest hutch. Measure enough tubing to run from the tank along the entire top line of hutches, plus an extra couple of feet in case you have to adjust the tank height. Fasten the line of tubing with small standoff clips to keep it away from nibbling rabbits. If you have only one level of hutches, you can put a drain valve at the end of the line. If you have a lower tier of hutches, run the tubing down to the first hutch on the end opposite the tank, then along the row of hutches, and put the drain valve at the end of the line.

Next, snip the tubing where you want to position each drinking valve, and insert the cut ends into a tee connector. Add 12" of tubing to the bottom of the tee, and attach the drinking valve and a valve clip.

Starter Kits

A handy way to start with a flexible tube system is with a starter kit, such as KW Cages' Nivek Starter Kit (see Resources), which includes all of the components you need. My favorite system is the Edstrom system. You can pick and choose the various components you want and purchase them separately. If you decide to go with the tubing system, be sure to use black plastic tubing. Algae can grow in clear tubing, causing blockages and malfunction of the system.

❊ AUTOMATIC WATERING SYSTEM ❊

A flexible tubing watering system is inexpensive and easy to install. Components are available from rabbitry suppliers.

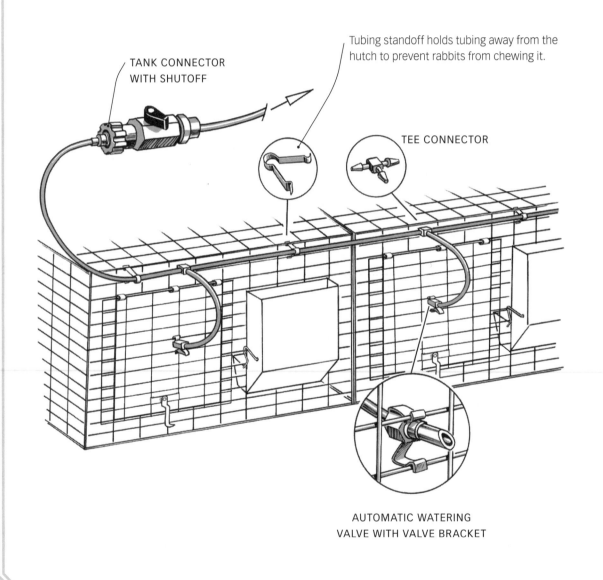

TANK CONNECTOR
WITH SHUTOFF

Tubing standoff holds tubing away from the hutch to prevent rabbits from chewing it.

TEE CONNECTOR

AUTOMATIC WATERING
VALVE WITH VALVE BRACKET

Repeat this at each hutch, and you're done installing the tubing.

If you will deal with freezing temperatures, you can do several things. If your rabbitry is not in an insulated and heated building, as many are not, the simplest way to deal with the cold is to drain the system. If it's freezing only at night, you can drain the system at the end of the day. During continued freezing temperatures, you can leave the system drained and use metal pans for watering by hand until things warm up. You don't have to worry about freezing temperatures breaking anything because it's all flexible plastic.

You can also try placing an immersion or bucket heater in the tank to warm the water there. If you leave the drain valve to the tank open, the heated water will move through the tubing. You can also use a small pump to recirculate the water back to the tank. Obviously these options require electricity, which is wonderful to have and really necessary for year-round production because you need some lighting during the winter. But if you don't have electricity, you can revert to draining the system and watering by hand during the time of freezing temperatures. (I have done that in my building in Vermont, where the temperature is below freezing from December to March. I'm 40 miles from the Canadian border. If you are farther south, and practically everybody is, you will probably have fewer freezing nights than I have.)

Rigid Pipe System

An alternative to flexible plastic tubing is a rigid pipe system utilizing ½"-diameter plastic pipe that can accommodate one or two heating cables inside, if desired. You need a hacksaw or a pipe cutter to cut the pipe, and you assemble the pipe, elbows, and tees with plastic pipe cement (solvent glue). Pipe saddles are required to hold the valves. A rigid pipe system is more expensive and more work to install than flexible plastic tubing, but it does make it easier to prevent the water from freezing.

You can run a single heating cable inside the tank and pipe and tie it to a thermostat so you can program when the heater comes

✳ FREEZE-X SYSTEM ✳

This watering system component includes a float valve for the automatic supply of piped-in water, and an immersion heater and pump to keep water from freezing.

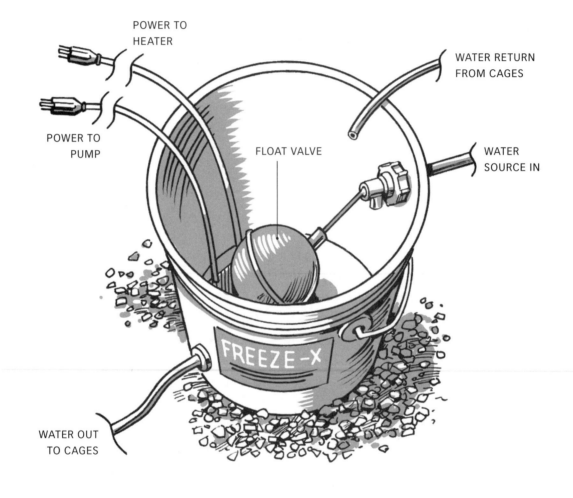

POWER TO
HEATER

POWER TO
PUMP

WATER RETURN
FROM CAGES

FLOAT VALVE

WATER
SOURCE IN

FREEZE-X

WATER OUT
TO CAGES

on. My New Jersey rabbitry had a rigid pipe that required two cables: one on a thermostat and one that I turned on manually when the temperature dropped below about 10°F. That kept the water heated until the temperature reached about 0°F. In Vermont it gets below zero regularly, and once the temperature really sinks, the rigid system freezes up. Fortunately, the pipe won't crack because it's plastic.

One of the drawbacks of a heating cable is that it cannot be lengthened or shortened. It is available in various lengths, but if you decide to expand your rabbitry, you can't add onto it. Typically, heating cables use 2¾ watts per foot. One cable performs well to about 20°F; two cables installed in the same pipe will protect to about 0°F. If you use two, it's a good idea to put one on a thermostat, which doesn't cost very much. One brand is the Thermocube, which can turn on a cable or pump automatically when the temperature drops to 35°F. The other cable can be turned on manually if it gets much colder.

Some rabbit raisers use the kind of heating cable that keeps ice from forming on roofs, wrapping it around the pipe. However, rabbits can chew on these cables if they have access, and much of the heat from the cable is lost heating up the outside air. By contrast, a heating cable inside the pipe is in direct contact with the water, so all of the heat goes into it.

It is quite a job to insert the heating cable into long lengths of rigid pipe, and it's best to do so while you're configuring the water system. It's also a good idea to drill the holes for the valves and saddles before inserting the cable so as not to puncture it, then install the cable into elbows and tees as you put them on, because it's tricky to get it around 90-degree angles. An electrician's fish tape is ideal for the job, but a long length of stiff wire works, too. You just tape the end of the cable to the end of the wire. I have even used a bamboo fly rod. Heating cables come with a wiring diagram and instruction sheet to guide the installation and help make the wiring connections.

If you want to avoid the difficulty of installing heating cable into a rigid plastic pipe system, you can use what's called the Freeze-X System (see box at left). It keeps the water moving through the low-pressure water lines of your automatic watering system, which is similar to leaving a tap in your home dripping on a cold night to prevent the water line from freezing. The system works very well when installed properly, but it does have limitations. If your cages are stacked or several lines are connected in a series, the workable length of pipe that one pump can accommodate will vary. Nothing, however, will prevent a line from freezing in single-digit temperatures and below if the wind is blowing directly on the pipe (which is certainly not good for your rabbits either). Insulation and windproofing play important roles in the ultimate effectiveness of the system.

The Best Nest Box

There is one more item that goes inside the doe hutch: the nest box. In yesteryear, when hutches had solid wood floors, you didn't need a nest box. The floor was bedded with lots of straw and the doe would gather some of it up, pile it in a corner, and build her nest there. For hutches with wire floors, however, you need a nest box. You know how much I like wire hutches. I also like wire nest boxes. And, if you order a roll of floor wire to make your own hutches, you will have plenty left over for making boxes.

Wood versus Metal

You could build a nest box using wood, and lots of people do, but unless you protect the edges with metal, the doe will eventually chew it up. That's not so terrible, especially if you have more wood. The more significant drawback of the wood box is that it's difficult to keep it sanitary. The doe and her youngsters will urinate in the box, the urine soaks into the wood, and no matter how much you wash and disinfect, you will never get it completely germ-free.

A metal nest box that is open at the top is a better choice. The closed ones tend to make things inside moist, which is bad news for baby rabbits. Rabbits that bump against the sides find them pretty cold in a northern winter. Open nest boxes make it easy to see inside year-round, they're easy to keep warm in winter with a warmer pad or light bulb, and they're a cinch to keep cool in the summer. Years ago, before there were wire nest boxes, rabbit raisers built "cooling baskets" of hardware cloth to keep litters in during the daytime. With the wire nest box you don't need a cooling basket; you simply use the box without the full liner, and it stays cool, ventilated, and dry.

If you have a wire nest box made of ½ × 1 floor wire, you can use a corrugated cardboard liner, which covers the floor and extends up the sides for each litter. You can buy these liners, but it's easy to make your own. A second wire floor goes on top of the cardboard to protect it, and shavings and straw are placed on top of that. The wire nest box is open. Some of the time, when the weather is warm, you just need some cardboard on the floor.

Flanges versus No Flanges

Flanges are pieces of galvanized sheet metal that cover the sharp edges of the top of the nest box and hold the liner in place. They provide a very smooth top edge, which is handy when rabbits are hopping in and out. This type of nest box is easier to make than the type without flanges, but it cannot be used as a subterranean nest and would need to be simply placed inside the hutch. The nest box without flanges can be used either way.

Installing a permanent subterranean nest below the hutch floor is a good option if you have only one tier of hutches. If you have two or more tiers, it would be difficult to do because you would need about a foot between the bottom of the top cage and the top of the one below. With a subterranean box, you don't keep a liner in the nest when it's not in use by the doe and litter.

You can buy wire nest boxes or you can make your own. My homemade boxes have lasted for about 40 years and are still as good as new. They are light so you can store them right on top of the hutch if you like.

Heating Tactics

Whatever type of nest box you make, you can warm it in very cold weather with a nest box warmer pad or a 25-watt lightbulb with an aluminum photoflood reflector underneath. A ceramic aquarium reptile heater bulb also works well. The warmer pads come in different sizes for standard nest boxes and heat up to 96°F when you put them under the second floor because they are insulated by the cardboard, shavings, and straw. The photoflood reflector can be punched for S-hooks and attached to the floor wire underneath the box.

Always remember to keep electric cords below hutch floors so they cannot be chewed and make sure any hay, straw, or wood shavings are away from heat sources. The electric cord goes through the floor wire to an extension cord. To protect against shock in the event of a short circuit, either the extension cord or the outlet it plugs into should be GFCI (ground-fault circuit-interrupter) protected.

In very cold weather, it's also a good idea to place an additional layer of cardboard or even some plastic foam, as well as an inch or so of shavings (for moisture absorption and insulation), between the two wire floors. I save the plastic foam trays from the packaging of meat and produce from the supermarket. You should also add additional shavings and plenty of straw above the second floor, filling the box to the top. The doe will burrow into it to make a nest and line it with fur from her own body. Insulation foam-board panels, cut to fit and placed on the top and sides of the hutch, will also help keep a new litter warm.

One more note: Never use a plastic cat bed or a cardboard box alone as a nest box. Rabbits chew plastic, and the doe will tip over a cardboard box if she doesn't chew it to pieces first. Either can spell disaster for the litter.

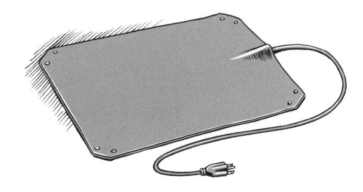

Nest box warmer pad: two layers of sheet metal sandwich a heating cable. It goes under the nest box.

⇨ *If you don't want a subterranean nest, the easiest nest box to make is one with flanges. The flanges will cover the sharp edges along the top of the nest box and clamp the corrugated liner in place. The assembled box measures 10″ × 18″ × 8″ and will work for all but giant breeds.*

MATERIALS

- 4 feet of ½″ × 1″ welded wire mesh, 14- or 16-gauge, 18 inches wide
- 28 J-clips
- 12″ × 17″ piece of 28-gauge galvanized sheet metal
- Corrugated cardboard

TOOLS

- Heavy-duty wire cutters
- Tin snips
- Work gloves
- J-clip pliers
- One piece of ¼″-thick × 2″ or 3″-wide wood lath, about 24″ long
- One ½″-diameter wooden dowel, about 24″ long
- Hammer
- One piece of 1×2 scrap lumber, at least 17″ long
- Utility knife or box cutter
- Screwdriver

Make the Nest Box

1. Cut a 10″ × 18″ piece of ½″ × 1″ wire mesh, using wire cutters, for the lower floor.

2. Cut an 8″ × 10″ piece of wire for the back section, and cut two 8″ × 18″ pieces for the side sections.

3. Cut the front section according to the size of your rabbits: For medium-size breeds, cut the front at 8″ × 10″; for smaller breeds, make the front 6″ × 10″. The lower front affords smaller rabbits easier passage.

4. Assemble the sections using J-clips. Place a clip every 4″, starting with the corners. Make sure the horizontal wires are on the inside of the hutch, where they will make neat, tight corners. After a little practice you may find that you need to give the J-clips a second squeeze with the pliers for a tight grip. When fastening the floor wire, make sure the ½" wires face toward the top of the nest box.

Make the Flanges

1. Cut the flange pieces from galvanized sheet metal (available in sheets of various sizes at hardware stores and home center stores), using tin snips and wearing gloves to protect your hands. Cut three 3" × 17" pieces for the sides, and cut two 3" × 9" pieces for the front and back.

2. Create a jig for bending the flanges by nailing a piece of wood lath to a work bench (or you can use a heavy plank set on the floor or a table).

3. Lay one of the flange pieces lengthwise over the lath, with 1" of its width on top of the lath and 2" hanging off the edge. Place a ½" dowel on the sheet metal at the edge of the lath and hammer a slight curve into the metal by striking the dowel along its length.

4. Place your thumbs on the metal, and use your fingers to pull the metal around the dowel. Lay the flange over a short piece of 1×2 (or similar scrap wood), and strike the flange with your hammer. Turn the piece over and continue striking until you have the desired shape.

5. Repeat the procedure to shape the remaining flanges.

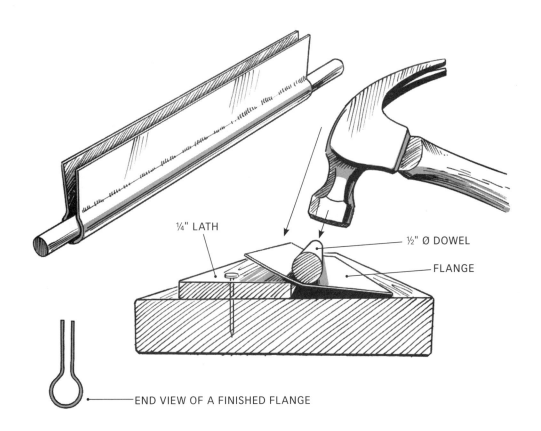

¼" LATH

½" Ø DOWEL

FLANGE

END VIEW OF A FINISHED FLANGE

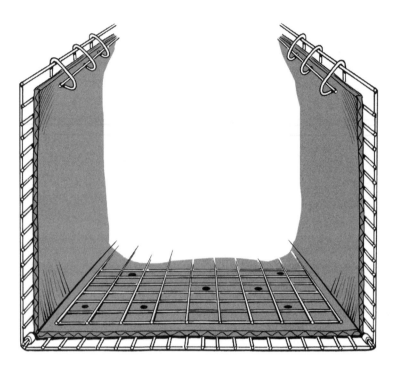

A "second floor" sits on top of the cardboard liner.

Fit the Cardboard

1. Cut the corrugated cardboard to fit the nest box, using a utility knife. Make the liner 1" smaller than the nest all the way around, and score lines for folding the cardboard into the box shape.

2. Fold the liner into the box shape and place it into the nest. Fit the flanges over the cardboard and wire to hold the cardboard in place and prevent chewing and scratching.

3. Cut out a "second floor" from ½" × 1" wire at 9" × 17" and place it inside the box over the cardboard bottom. Use a screwdriver to poke several drainage holes through the cardboard bottom.

Note: In warm weather, you will need cardboard only on the floor. In this case, twist lengths of wire around the flanges and through the wire box sides, front and back, to secure the flanges. You don't want them to come loose and injure the doe or her litter when they hop in and out of the box.

⟹ *This box is constructed much like the box with flanges, but here you make the sides, back, and front higher and turn the edges in at the top, eliminating the need for metal flanges. The assembled box measures 10" × 18" × 8". You can easily modify this design to create a subterranean nest (see Making a Subterranean Nest Box, on page 63).*

MATERIALS

- 4 feet of ½" × 1" welded wire mesh, 14- or 16-gauge, 18" wide
- 28 J-clips
- Corrugated cardboard

TOOLS

- Heavy-duty wire cutters
- Slip-joint pliers
- J-clip pliers
- Utility knife or box cutter
- Screwdriver

Make the Nest Box

1. Cut a 10" × 18" piece of ½" × 1" wire mesh, using wire cutters, for the lower floor.

2. Cut a 9" × 11" piece of wire for the back section, and cut two 9" × 19" pieces for the side sections.

3. Use slip-joint pliers to bend the top 1" of the side and back sections inward, leaving enough of a gap to slip in the corrugated cardboard. This procedure also provides a smooth edge.

4. Cut the front section according to the size of your rabbits: For medium-size breeds, cut the front at 10" × 12"; for smaller breeds, make the front 8" × 12". The lower front affords smaller rabbits easier passage.

5. Bend the top 1" of the front piece inward, then bend 1" on each side of the front to a 90-degree angle toward the rear (see illustration below).

6. Assemble the sections using J-clips. Place a clip every 4", starting with the corners. Make sure the vertical wires are on the outside of the hutch. That way the horizontal wires are on the inside where they will make neat, tight corners. You may need to squeeze the J-clips a second time with the pliers for a tight grip. When attaching the floor, make sure the curve bows toward the top of the nest box.

Fit the Cardboard

1. Cut the corrugated cardboard to fit the nest box, using a utility knife. Make the liner 1" smaller than the nest on each side, and score lines for folding the cardboard into the box shape.

2. Fold the liner into the box shape and place it into the wire nest.

3. Cut out a "second floor" from ½ × 1 wire at 9" × 17" and place it inside the box over the cardboard bottom. Use a screwdriver to poke several drainage holes through the cardboard bottom.

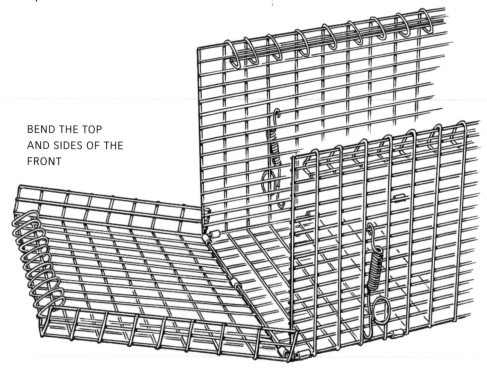

BEND THE TOP
AND SIDES OF THE
FRONT

Making a Subterranean Nest Box

When you J-clip the box together, clip only the bottom of the front and add springs with an S-hook to each side about halfway up. This allows you to swing the front down to examine and handle the litter as needed, then hook it back up. Cut an opening in the floor of the hutch at 10" × 18" and clip or ring the top of the box to the opening. When the nest is not in use, it simply becomes part of the hutch floor. Your hutch now has a sunken living room. Remove the liner, of course.

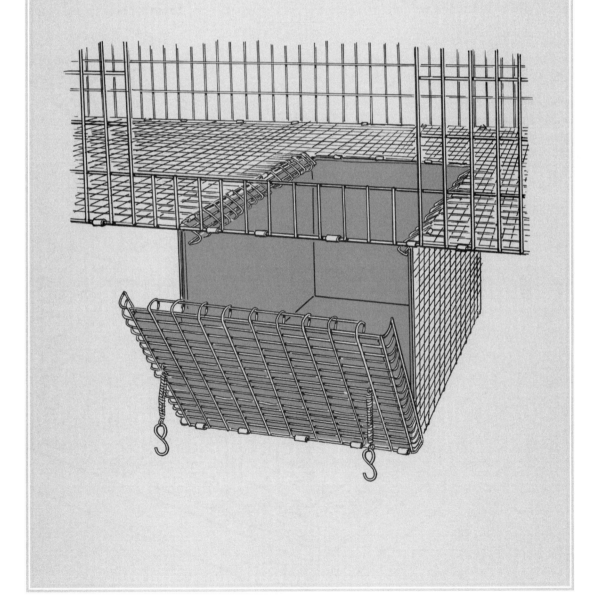

Building Your Own Carrying Cage

Sooner or later you will want to transport your rabbits to market, to a show, or just to take them out of the hutch to give it a cleaning. And just as with hutches and nest boxes, an all-wire carrier is the best way to go. It's light, well ventilated, and indestructible, and after building hutches and nest boxes, you will find it's a cinch to make.

The floor of the cage needs to be large enough for the rabbit to stretch out. For short hauls, the height can be only 8" to 12". For long trips, which might require feeding and watering along the way, a height up to 18 inches, depending on the size of the rabbit inside, is better. As with hutches, you can use 1" × 2" wire everywhere except the floor, which must be ½" × 1" wire. I've made several carrying cages

When making a carrying cage, place the ½" × 1" floor wire 2 inches from the bottom of the sides, to allow for clearance above the bottom of the pan. You can find a pan in a housewares department and build the carrier to fit it.

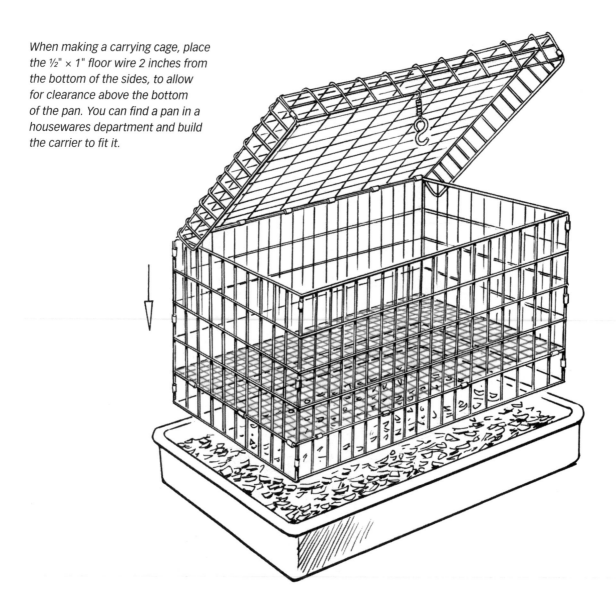

that I use at shows exclusively out of floor wire. The smaller mesh prevents small children from poking their fingers inside.

To provide clearance underneath the cage for drainage and droppings, raise the floor by clipping or ringing it 1" or 2" from the bottom edges of the sides. You can find plastic trays or pans in housewares departments and build your carriers to fit them. Put 1" or so of shavings in the pan and attach it to the carrier with short lengths of wire or springs and S-hooks.

Cut the top of the carrier 1" or 2" larger than the floor, and bend the edges to fit over the sides. Hinge it at the back with J-clips or hog rings. You can make a latch from a dog-leash snap fastener or a spring and an S-hook, as shown for the top-opening hutch (see page 37).

Of course, you can also buy carriers premade, including some with two or more compartments. But if you build your own and measure accurately, you can make it just the right size. Again, the welded wire grid ensures everything comes out square and true. Chances are, it won't be long before you consider building them to sell to other rabbit raisers.

After building several carriers with one or two compartments, I decided to build some with four and even five such compartments, which made for fewer trips to the showroom when I exhibited my stock. You might custom-build carriers to fit the space in the cargo area of your vehicle.

BENNETT RABBITRY #4
Shed Living

Rabbitry number four materialized when I decided to move the operation into a 10 × 12-foot shed attached to the house. The shed was also home to a small workbench, bikes, the barbecue, a lawnmower, and outdoor furniture during the winter. The impetus for the move came from an influx of small children in the neighborhood who delighted in poking sticks into the rabbit hutches and otherwise harassing the animals.

I built six all-wire hutches and fastened them three-high to 2×2 legs. I built this setup before equipment companies began to offer steel leg kits that I could have used instead of the 2×2 lumber. Another method would be to use plastic plumbing pipe.

Sloping galvanized dropping trays drained into 3-foot aluminum eave troughs (gutters) with end caps, which sat on brackets at the rear of each cage. Each cage also had a self-feeder. I used water bottles in warm weather and crocks in the winter. The two units rolled on casters on the concrete floor. Two windows admitted plenty of light and fresh air. My rosebushes, azaleas, and pachysandra beds were treated regularly to the contents of the gutters, and boy did they flourish (if the garden had been sunny enough for vegetables, they probably would have thrived, too).

So You Want Mini Satins? *Get Them from Zoom*

Zoom Rabbitry
Jim Krahulec

LOCATION: O'Fallon, Illinois

AVERAGE RABBIT POPULATION: 100

GOAL: Show and breeding stock

BREEDS: Mini Satin, Lionhead

JIM KRAHULEC lives just 15 miles from the St. Louis arch, the Gateway to the West. Inside his Zoom Rabbitry, comprising 112 wire cages inside a 20 × 40-foot former chicken coop, he developed a new breed of rabbits: the Mini Satin. That is known by rabbit breeders from coast to coast, but not by most of his neighbors.

"The rabbitry building was built in the 1940s by the original owners of my place, which then included 200 acres of farmland. Over the years they sold some of it for housing subdivisions on four sides. I bought the remaining six acres with the farmhouse and four outbuildings in 1995.

"So I have the best of both worlds: the privacy of a six-acre farm with the advantages of living right in the midst of the city. Only one of my neighbors even knows I have rabbits. The coop is shaded in the middle of a grove of hackberry trees, which some people call ironwood. Ivy grows like crazy on all four sides of the coop. The neighbors are nowhere near the coop or my house."

Inside, Jim's hutches are stacked four- and six-high. There is one row against each wall and two more rows that are back-to-back in the middle. When the temperature plummets below 10°F, Jim runs two electric space heaters. In the summer, he runs three window fans and a large squirrel-cage fan. His 8-foot fluorescent shop lights run on a timer from 6 A.M. to midnight, year-round.

Jim started a small watering system about 10 years ago but did not like it because the water froze too often in the winter. Now he uses 32-ounce water bottles for all hutches during the spring, summer, and fall, and he uses crocks during the winter. Because he is on city water, he fills 32 1-gallon plastic milk jugs and places them on shelves to let the chlorine and fluoride evaporate before giving the water to his rabbits. He uses about eight jugs per day to fill the water bottles. Jim says feeding time is enjoyable because it gives him exercise and a time to relax.

Jim and his late sister Carol had rabbits as children, and Carol had a large rabbitry as an adult before Jim did. Jim didn't get into the business for meat or fur sales, and he didn't sell manure or worms, either. His primary purpose was, and continues to be, to breed quality small rabbits for shows and sales. The manure goes into his garden and is free to many gardener friends.

Jim started his rabbitry in Los Angeles in 1982 with Mini Lops. The man who got him started knew he had been in the Air Force and always called him "Zoom," hence the name of his rabbitry. When he moved to Illinois, he added Holland Lops, and over the next several years he won many Best of Breed awards with both breeds. Then he started raising American Fuzzy Lops, those little rabbits that resemble Angoras, before they were recognized as a breed. They were a big success at shows as well. But that wasn't enough for Jim.

"Since Mini Rex had become so popular [now the most popular breed at ARBA (American Rabbit Breeders Association) shows], I had to have them," Jim said. "Before long I realized I had too many breeds to do them all justice, so I sold out

---

Jim didn't get into the business for meat or fur sales, and he didn't sell manure or worms, either. His primary purpose was, and continues to be, to breed quality small rabbits for shows and sales.

---

the Mini Lops and decided that if the Mini Rex could be created, I would invent the Mini Satin."

He started in 1993 with some unrecognized Satinette does and some undersized Satin bucks obtained from three friends. He then wrote the Mini Satin breed standard and presented five colors to the ARBA standards committee. After a few bumps in the road, the breed was recognized by ARBA in 2005, but only in white. Four more colors have been accepted since then, and about a

dozen other color varieties currently are in various stages of acceptance.

It takes a lot of rabbits to create a new breed, and Jim had more than 250 at the start of 1998. He even added Dutch and Lionheads to the Zoom operation. But the numbers were getting too large, so he decided to cut down on the herd and at last count he had about 100 Mini Satins (including 17 color varieties!) and 24 Lionheads.

CHAPTER FOUR

Sheltering a Backyard Rabbitry

Now you have your hutches and have equipped them to feed and water your rabbits and even provide for litters, so let's examine some of your options if you plan to use these hutches outdoors. If you don't already have a building in your backyard to house your rabbits, you'll need to determine the best place to put one. You'll also need a structure that comfortably and safely accommodates the herd, and preferably yourself.

This chapter includes ideas, plans, and illustrations for some simple three-wall structures, as well as some larger closed buildings. For more ideas and plans, I recommend two books: *How to Build Animal Housing* and *Building Small Barns, Sheds & Shelters* (see Resources).

In my lifetime I have had 16 different backyards (thanks to being an Army brat and having an adventurous mother), so I have some familiarity with backyard landscapes ranging from Vermont to California. Since the days of my first ramshackle rabbitry on Mendon Mountain, I've been sizing up backyards to see how rabbits would fare there.

PROVIDING CREATURE COMFORTS. The most common kind of rabbitry I've seen over the years consists of one or two hutches standing on some kind of legs in a backyard. The illustration on the next page shows four hutches hanging under a slanted roof supported by posts. This simple setup lets manure fall to the ground unobstructed, and it works well if it's protected from the elements with a plywood front, back, and side panels, and if it's fenced in against predators. But you have to remember that when keeping rabbits you have to visit them every day, at least once, in all kinds of weather. Yes, every single day, no matter what.

❋ SIMPLE BACKYARD RABBITRY ❋

A structure like this is just a "table" with a slanted top (roof). It can be built to accommodate any number of hutches, and, depending on your climate and weather conditions, you may enclose it with weather curtains or panels.

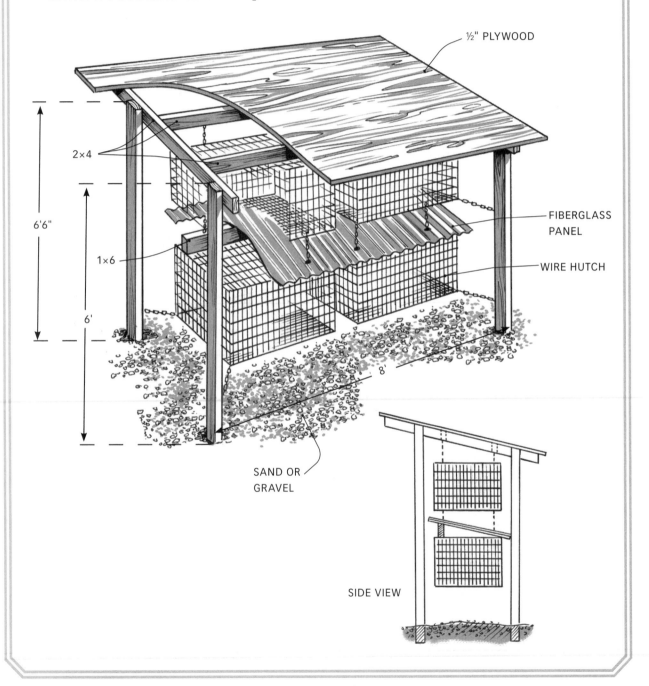

½" PLYWOOD

2×4

6'6"

1×6

6'

FIBERGLASS PANEL

WIRE HUTCH

8'

SAND OR GRAVEL

SIDE VIEW

Depending on your climate, there may be quite a few times when you won't relish the idea of going out to feed and water the herd, not to mention for breeding, nest box preparations, weaning, culling, manure removal, record keeping, and regular maintenance. Pouring rain, deep snow, wild wind, bitter cold, and burning sun come to mind. So I recommend placing your hutch or hutches inside something that will protect *you* from the elements as well as your herd. This can also provide another kind of security: Because your rabbits will look very good where you put them, or rarely be seen at all, you'll protect the enterprise from humans who may not appreciate rabbits as much as you do.

CHOOSING A LOCATION. You should locate your building fairly close to the house because you will go there daily. Behind your garage might be a good place. Next to a vegetable garden is a great place because rabbit manure will pep up your compost heap better than anything I know. In fact, if you're going to raise rabbits and don't have a garden, you may want to consider putting one in because the manure is wonderful. I don't think I could garden without it (my 8-foot-high tomato plants are prime evidence). If you do have a vegetable garden and it is fenced in, place your rabbitry inside the fence to keep it safe from intruders. I wouldn't put my rabbits next to a play area for young children. The laughter and activity on a swing set, jungle gym, or even a sandbox might delight the kiddies but keep the rabbits on edge.

FOR BUILDING AND ZONING REQUIREMENTS. Regardless of the type of building you plan to have, the first thing to do is check your city building codes and zoning laws. Municipal rules may govern the size and type of structure that's allowed, as well as its proximity to your property lines. Often small buildings and those with "nonpermanent" foundations (landscape timbers, concrete blocks, railroad ties, etc.) are subject to fewer restrictions than large structures with concrete slab foundations. Permits may also be required for running electricity or water service out to the building.

Simple Three-Wall Structures

It's a simple matter to put a roof over your rabbit hutches to protect them from the elements, and it's really no more difficult to protect the rabbit keeper at the same time, both in terms of the work involved and the materials required. A proper sheltering structure not only looks better than an assemblage of several covered hutches, it can also look like anyone else's ordinary outbuilding for lawnmowers, garden tools, and seasonal storage. Following are three basic designs for three-wall structures, along with an overview on how to construct each.

A structure like this is easy to build and it protects the rabbit raiser as well as the rabbits.

Pole Building

My favorite structure for keeping rabbits secure and protecting them and their keeper from the elements is based on what farmers call a machine shed, or a loafing shed for large animals. If you hike the Appalachian Trail, you might recognize it as something like an Adirondack lean-to. It also reminds me of a baseball dugout, and I have seen a lot of bus stop shelters with a similar design. The structure is open to the elements unless closed up in times of severe weather, you don't have to do anything special to ventilate it, and heat is not really an option.

With a building like this, you can stack your hutches inside, using pans underneath, or hang them from the rafters. With electricity you could put lighting overhead. Rabbitries tend to grow in time, and this setup would allow for expansion. You can make it longer or face two of them 3 or

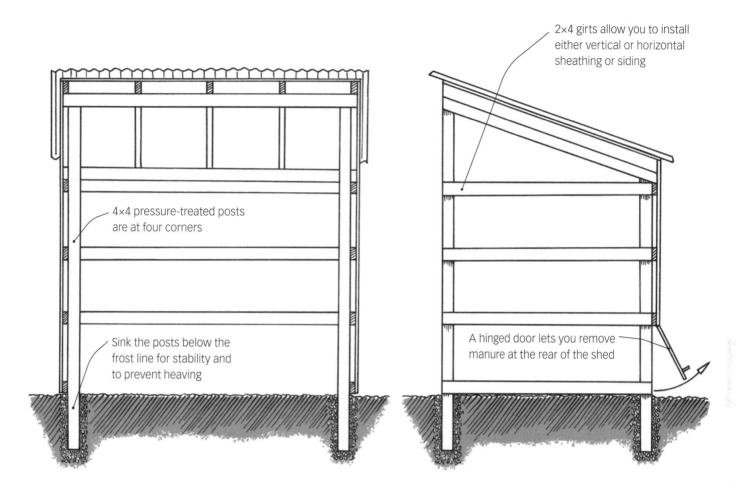

2×4 girts allow you to install either vertical or horizontal sheathing or siding

4×4 pressure-treated posts are at four corners

Sink the posts below the frost line for stability and to prevent heaving

A hinged door lets you remove manure at the rear of the shed

This shed (similar to the one shown on the facing page) is inexpensive and simple to construct.

4 feet apart and put a roof over the aisle in the manner of a monitor barn. You can also partition a section and make a closet for storing hay, feed, nest boxes, and tools.

A 6 × 8-foot building could accommodate up to nine 30" × 36" hutches stacked three tiers high, and even more if you raise a small breed or want to devote some hutches to bucks and growing stock. That would leave 3½ feet in front of the hutches to shelter you, not counting the roof overhang. You might face the shelter to the east to capture the morning sun or to the south for rays all day. I would put it parallel with or at right angles to other buildings, such as the house or garage, and to the property line. Be sure to check up on municipal rules about setback footage from your property lines (see page 11, Neighbors and Laws).

Front "Wall" Options

There are a number of ways to shield a three-wall shed from the elements. You can add storm curtains made from canvas or vinyl tarps or clear plastic sheeting, or install storm doors on the front for more permanence. In the summer, you can add exterior Roman shades to let fresh air through while blocking most of the sun's heat; you just raise or lower them with a drawstring. Roll-up solar shades also block hot sun and glare and let cooling breezes through. You can raise or lower them to any level with a pull chain. Magnetic screen curtains can help keep flies away. Sliding louvered doors designed for closets (available at home centers) make handy doors. Shade cloth used in nurseries to protect plants will let air through but keep the sun outside. You might even consider adding an awning to the front to reduce heat and glare inside as well as shield against rain and snow.

Canvas or vinyl tarps, or even clear plastic sheeting, will protect your rabbits during storms, high winds, or severe cold.

Sliding or bifold louvered closet doors are another option to protect against the elements — and to protect against predators at night.

An awning is still another way to keep out inclement weather and the hot sun in summer. It can be lowered to cover the front of the structure in case of strong winds or extreme cold.

CONSTRUCTING A POLE BUILDING

For a 6 × 8-foot shed, dig post holes at the four corners to a depth of at least 18" (based on the depth of the frost line in your area). Fill each hole with 6" of tamped gravel or 4" to 6" of concrete (let this set for at least a day). Insert 4×4 pressure-treated posts into the holes, and brace them with cross bracing so they are perfectly plumb. Secure the posts with alternating layers of tamped gravel and soil. Alternatively, you can secure the posts with steel fixtures called "post-ups" driven into the ground. The two rear posts should extend 6 feet above the ground, and the two front posts should extend 8 feet. It's usually easiest to cut the posts to exact length after they're in place.

Nail double 2×4 plates on all four sides and install single 2×4 horizontal girts (not vertical studs), spaced 2 feet apart, below the plates on the sides. Nail only two girts on the back, because you are going to leave the bottom 2 feet open for manure removal (see illustration on page 73).

Install 2×4 rafters spaced 2 feet apart. You might sheathe the structure with plywood plus siding and roofing to match your house. Otherwise, you can use board-and-batten vertical siding, cedar shakes, corrugated fiberglass roofing panels, or whatever suits you.

It's a good idea to panel the inside with a very smooth, water-resistant material, such as cement board or vinyl-coated tile boards, or you can use the corrugated fiberglass panels, which are waterproof and easy to clean. Insulation is not necessary in such an open structure, except possibly for the roof if there is little or no shade afforded by nearby trees. You might put windows or shutters on each end, especially if you enclose the front during inclement or freezing weather.

You can simply rake the manure out the back, over a tamped earth floor, and shovel it into a wheelbarrow or cart and take it to your compost heap or directly to your garden. A hard-surface walkway in front adds a finished look and prevents a muddy approach; it can be poured concrete, gravel, concrete or brick pavers, or flagstones. Laying down another hard surface behind the shed facilitates snow removal when it's time to clear out manure during the winter.

If you raise small rabbits with 2-foot-deep hutches, or if you just want a smaller structure, you could build a 4 × 8-foot or 4 × 6-foot building that would still provide a roof over your head. Alternatively, you could build two 4-foot-deep structures facing each other with an aisle between, then add a roof over the aisle.

An arbor over the entire structure is a nice addition to consider. Roses or climbing flowers, such as morning glories or wisteria, can provide beauty, shade, and even camouflage. Grapes or squash would do the same plus give you something to eat. Roses and squash, in particular, would love the manure you remove from the hutches.

Roll-up tarps of canvas or vinyl can protect your stock from driving rain or snow.

Basic Shelter with Angle Iron Framework

A good friend of mine built a basic slanted-roof shelter using perforated-steel angle iron, a very versatile framing member. The description of this comes from my book *Storey's Guide to Raising Rabbits* (see Resources). The structure requires only nuts and bolts for assembly — it goes together like a child's construction set — and you can find the materials at most home centers and hardware stores. The metal framing not only supports the lean-to structure, it's also strong enough and handy for suspending the hutches (with the rabbits, of course). Apart from the sheathing and roofing, no wood was used.

CONSTRUCTING AN ANGLE IRON SHELTER

Use the slots or holes as guides to measure the angle iron. Cut all the angle iron pieces, using a hacksaw or reciprocating saw, then assemble the frame with hex nuts and bolts, tightening them with a hex wrench and nut driver. It's easier if a friend helps with the assembly.

To anchor the frame, bolt it to concrete footings, pressure-treated wooden members, or railroad ties, which you might want to bury underground or anchor with rebar stakes or other means. If high winds are not a problem, merely setting the frame on bricks or concrete patio blocks can suffice. Clad the three walls with exterior-grade plywood bolted directly to the frame. (As an alternative to plywood, see Gabriel Roper's profile on page 84 for information on using corrugated fiberglass or PVC plastic panels for cladding.)

Build the roof with corrugated fiberglass roofing panels or with a layer of plywood finished with shingles or rolled roofing to ensure waterproofing. Be sure to provide an overhang in the front, to protect yourself from rain and to keep direct sunlight off the animals. Paint the outside of the structure, and you're done!

One of the nicest things about the angle iron is that you can also easily add extensions as your rabbitry grows, by simply bolting on more sections to the existing ones. On the other hand, if you ever decide to house your rabbits somewhere else or you give up keeping them, the whole structure can be taken apart and the angle iron used for shelving in the garage or basement, or even for a workbench frame.

ANGLE IRON SHELTER PLANS

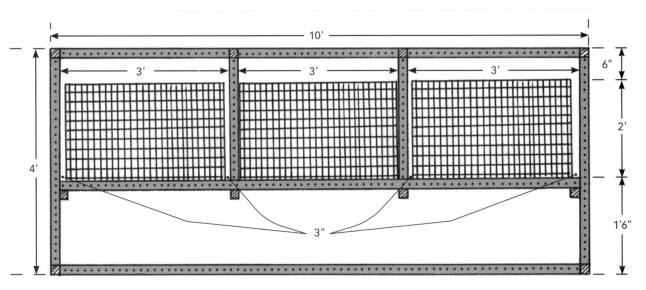

TOP VIEW OF FRAME AND HUTCHES

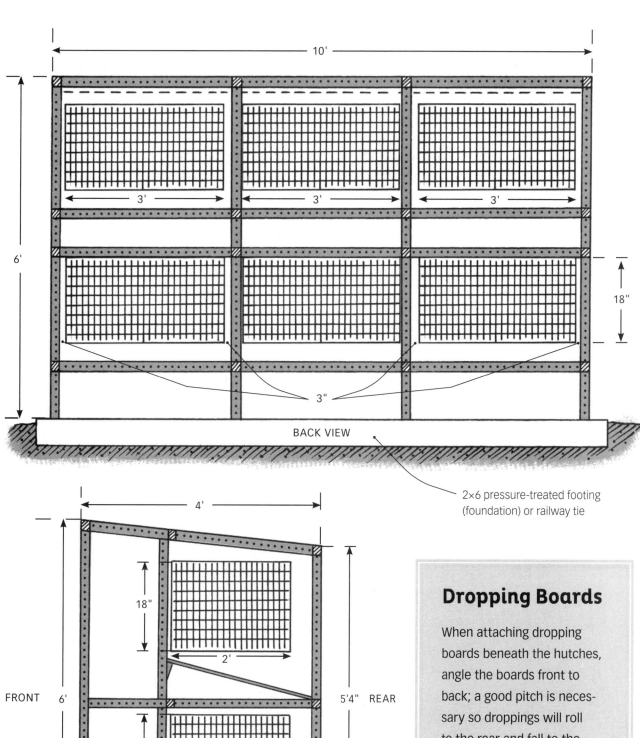

10'

6'

3'

3'

3'

18"

3"

BACK VIEW

2×6 pressure-treated footing
(foundation) or railway tie

4'

18"

2'

FRONT

6'

18"

2'

5'4" REAR

SIDE VIEW

Dropping Boards

When attaching dropping boards beneath the hutches, angle the boards front to back; a good pitch is necessary so droppings will roll to the rear and fall to the ground. Leave enough room in front between the cage bottom and the droppings board so you can use a scraper or hoe occasionally on the droppings board.

Garage Lean-to

One classic alternative to a free-standing building is a lean-to attached to a garage. This pole structure gets a pedestrian door at one end, or you can access the lean-to through a door in the garage wall. If your garage is attached to your house, you could care for your herd without ever going outdoors. And if the garage is heated, you could leave the door open to allow warm air in (but not when you are starting the car because of the threat of carbon monoxide). Such an arrangement reminds me of some New England farmsteads, where a garage or shed is attached to the back of the house and a barn is attached to the garage or shed.

When deciding where to locate the lean-to, consider sun, possible shade, and prevailing winds. An arbor over this shed or a trellis planted with climbing vegetation can provide shade if necessary.

Note that the manure cleanout panel is hinged low on the side of the structure.

R. L. Wilson's "hoe house" in winter

Gabriel Roper's wire hutch

A Tour of Rabbitries (Rabbit Real Estate)

The four different rabbit raisers in the following pages demonstrate how the recommended all-wire hutch can be used in strikingly different ways, depending on individual needs and goals.

Dan Brink's big outdoor hutches

Sarah Ouellette with her stacked hutches in the former milk house

Roper's perforated steel shed

V-shaped hay rack separates and serves two hutch compartments in the Rare Hare Barn

Debbie Vigue's environmentally controlled rabbitry

The Rare Hare Barn
Callene & Eric Rapp

LOCATION: Leon, Kansas

AVERAGE RABBIT POPULATION: 160 adult breeding stock; about 800 with litters and fryers

GOAL: Raise meat rabbits sustainably, humanely, and profitably while conserving rare and endangered heritage breeds

BREEDS: American Chinchilla, Silver Fox, Crème d'Argent, Blanc de Hotot, American (Blue and White)

CALLENE AND ERIC RAPP run The Rare Hare Barn in Leon, Kansas, specializing in rabbit meat processed in a USDA-inspected slaughterhouse and shipped throughout the country by Fed Ex. You can order fryers fresh, frozen, or smoked, plus sausage, "bratwurst," and liver.

In addition, you can purchase breeding stock of several breeds. You can see the animals in person at the Sedgwick County Zoo, the Oklahoma City Zoo, or Tilly Foster Farm in New York. Can't get to any of those places? Check out their website (see Resources).

Callene is relatively new to rabbit raising but is a veteran zookeeper, so she knows a lot about animals. The county zoo maintains the largest collection of rare breeds of livestock in the country. Eric is a third-generation farmer who helped his grandfather raise rabbits and now operates The Rare Hare Barn full-time.

They operate their "barn" out of three carports fitted with wire hutches. The carports seem ideal. They were quick and easy to erect, were relatively inexpensive, afford excellent ventilation, and give the rabbits a good view of the Kansas countryside. To keep things cool in the summer they use misting nozzles and a swamp cooler, which is a water-soaked pad. In nasty weather they lower curtains around the carports.

The Rapps chose breeds they felt were uncommon to ensure their numbers didn't further

Callene holds a Blanc de Hotot; Eric has a Chinchilla.

A Blanc de Hotot doe and young American White sit in a kindling barn. The storm curtain is rolled up above the window.

decrease. The success of the Rapps' business proves once again that domesticated rabbits owe their continued existence to the fact that they make a delectable dish. I know that for sure because I have recently raised some of these breeds for my family's table. Probably the rarest of the Rapps' rabbits, the American Chinchilla, is known for a long tasty loin, almost as long as those on the giant breeds, but the American doesn't have the drawback of requiring more hutch space and feed. The Rapps also like the various colors of their

Clockwise from top left: American Chinchilla doe and litter; breeding and kindling barn; Callene feeding in doe barn; and developer barn.

chosen breeds, which they find more interesting than plain white rabbits.

With an excellent approach to marketing and the latest equipment, the Rapps are running a successful rabbit meat business right in the heart of beef cattle country.

Gabriel Roper

LOCATION: Bedford, Indiana

AVERAGE RABBIT POPULATION: 30

GOAL: Show and breeding stock

BREED: Champagne d'Argent and Broken New Zealand

IT'S A GREAT SOURCE of satisfaction when somebody follows your advice and succeeds, so I popped a few buttons on my vest when I heard from Gabriel Roper, a young nurse who lives in Bedford, Indiana.

Gabriel told me that he and his wife, Bethany, wanted to farm for a living some day but thought it would be wise to begin small with a few rabbits, so he bought *Storey's Guide to Raising Rabbits* and studied it carefully.

"It wasn't long," he told me, "before I started building my own custom rabbitry using suggestions in the book." The basis of his rabbitry was a plan for an angle iron shed that my former rabbit-raising partner, John Dack, built in New Jersey many years ago (see page 78). The shed is located about 40 feet behind his home and about 100 yards away from his nearest neighbor.

"Angle iron works very well," Gabriel said. "It's more expensive up-front than lumber but it will last us for years. Maintenance is easy. I simply spray it down with my garden hose. The corrugated fiberglass roof panels under the top row of cages are angled to allow waste to simply roll out the back of the rabbitry. I cut an extra piece of the roofing material and bolted it to an old broom handle with U-bolts to make a scraper/pusher, which makes maintenance even easier.

"We started in the winter. Changing frozen water crocks twice a day was very time-consuming. Pretty soon it became obvious that automatic watering would be a wise investment."

Gabriel holds his pet Champagne D'Argent.

Children Alaysa (left) and Reuben enjoy playing with baby rabbits. Frequent handling keeps the rabbits calm.

That's when Gabriel asked me to help him design a watering system, and I recommended the Edstrom components, which were supplied by Bass Equipment Company (see Resources).

"I paid $4 for each of my six crocks, but the $70 I spent on the watering system is worth every penny. I can devote more of my time to managing my herd rather than rinsing crocks. This rabbitry is a good design, and it can be expanded at any time or taken apart and moved to a new location. That was a critical point for us, because we rent right now but plan to buy land in the future. We raise Champagne

Clockwise from top left: A five-gallon bucket serves as a holding tank for the watering system and hangs high to provide proper water pressure; Gabriel's shed; daughter Alaysa watches Cally in her nest box; Bethany Roper's pet Broken New Zealand "Cally."

D'Argents and Broken New Zealands, a new variety, for breeding and meat, and we hope this will turn into a full-time business for us. To make our dream a reality, it was important that we constructed a rabbitry that was efficient to manage, durable, and flexible. I highly recommend this design for those reasons," he says.

Well, so do I (I mean, I already did).

Bobbi Daniels

LOCATION: Sitka, Alaska

AVERAGE RABBIT POPULATION: 5

GOAL: Small-scale fiber production from pampered rabbits

BREED: French Angora

I HAVE NEVER RAISED ANGORAS. I think they are beautiful and I know their wool is amazingly warm and soft. But from what I have learned, they have special requirements for feed, grooming, and housing, and I have never been willing to give them the time and effort they require. However, I know quite a few people who do raise them. One is Bobbi Daniels, who lives in Alaska. Here's what she has to say about hutches for Angoras.

"Wire cages are basically a necessity. Angoras simply cannot live in any sort of bedding, especially shavings. It is not just that it ruins the fiber for spinning. The damp shavings mat into their hair and they are continually damp with urine in all the hair that touches the floor, and I cannot begin to describe to you what a mess it is — to the point of being a health issue for them. It just simply doesn't work any way except with a wire floor.

"Also, they are passing lots of hair in their droppings, which is a good thing. And the long hair tends to create clusters of manure pellets — affectionately called 'strings of pearls' by Angora owners — that don't fall through the ½ × 1 mesh wire. For that reason, I use 1 × 1 wire on the floor and supply them with cardboard mats to sit on. They do not mess on the mats, and ripping mats apart gives them something to do. The ½ × 1 mesh

Angoras are often sheared but Bobbi plucks one here. Angoras shed or release their coats every 12–14 weeks.

It's important to give Angoras high-quality feed pellets to foster their extreme hair growth.

ends up with poop piled in the corner, just like it would be on a solid floor, and then it takes a propane torch to clean it out because all that long hair is wrapped around the wire. I have never had a problem with 1 × 1 for the floor. The cages I build for the expectant moms do have the ½ × 1 mesh because the baby feet, and sometimes the new baby, can go right through the 1 × 1.

Bobbi points out that when it comes to nest box bedding, straw seems to be satisfactory, both for the babies and the mom.

TOP: Bobbie uses plastic zip ties from the hardware store, instead of J-clips, to join wire hutch panels. She says they work just fine.

BOTTOM LEFT: Stay Puft (that's its name) noses right up the camera.

BOTTOM RIGHT: Angora fiber comes right off the rabbit ready to spin. Few things, Bobbi says, can top spinning with the rabbit right on your lap.

Kerry Hogan

LOCATION: Minnesota

AVERAGE RABBIT POPULATION: 12

GOAL: Home meat supply

BREEDS: Champagne d'Argent and Silver Fox

WITH A CITY LOT not far from Minneapolis, Minnesota, Kerry Hogan elected to build a rabbit shed next to his garage, which is behind his home. Its dimensions puzzled me until Kerry explained that you don't need a permit for a freestanding building smaller than 10 × 12 feet. His is about 7½ × 9½ feet, has neither a front nor

Louis, the Champagne D'Argent buck, gets his head scratched by Kerry.

Two tiers of wire hutches hang inside the shed.

a back wall, and shelters nine all-wire hutches stacked two and three tiers high. The floor is dirt.

Kerry raises Champagnes and Silver Foxes for a home meat supply. The hutches hang from above on wires, and Kerry uses corrugated PVC dropping "boards" cut from 26" × 96" panels. A cabinetmaker who builds kitchens and furniture with solid woods and fancy veneers, he knows his tools and techniques and brought them to bear when he hung the hutches, as you can see from the photos that his wife, Penelope, shot for him.

Kerry rakes off the front-slanting dropping boards daily, and the manure goes to a compost bin that's 20 feet away, near his 24 × 33-foot vegetable garden. When he opens the bin in the spring, he says, worms "waterfall out of it like Niagara."

Only 8" away from the hutches, the vinyl siding on the garage serves as a back wall. The space provides good air circulation. Storm curtains can be pulled down at both the front and rear of the shed. Because it gets extremely warm in Minnesota (as well as famously cold), Kerry has an air conditioner for the window on one side.

He removes the hutches periodically to clean behind them and says he plans to add housewrap or plastic to the interior walls. His wintertime routine is to move some of the hutches into his basement woodworking shop, stacking them three high, and using dropping pans. Open windows provide adequate ventilation in Minnesota, where the wintertime air is fresh, sometimes even impudent.

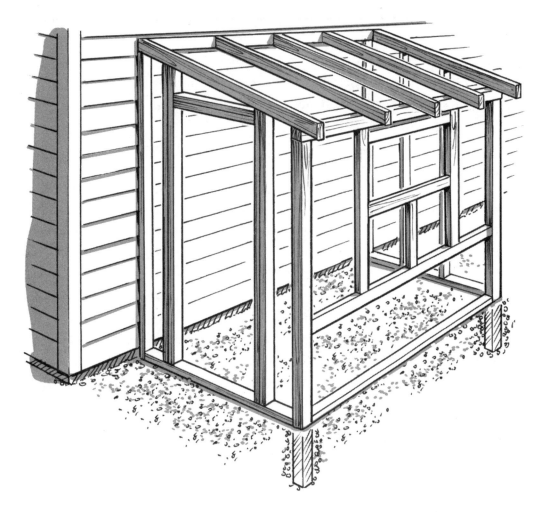

Shown here is the easy pole-construction framework for the garage lean-to on the facing page.

CONSTRUCTING A LEAN-TO

Set two posts as for the pole building (page 73), placing them 6 feet away from the back or side of the garage. Build a shed-style roof scabbed to a ledger board that's lag-screwed or spiked to the garage wall's framing, or extend the gable roof by adding extensions to the existing rafters, first removing fascia and soffit boards. A good overhang and a rain gutter will shield a window near the top on the long side wall of the structure.

You can finish the shed to match your garage. For convenience and additional ventilation, attach the siding to a hinged panel comprising the bottom 2 feet of the side wall. Open the panel for manure removal and ventilation, and close it against rain, wind, or snow. Outfit the shed with hutches in the manner of the free-standing structures, either by hanging them from the rafters or using stacking legs with dropping pans.

Closed Buildings

If you decide to house your herd in a closed building, as opposed to a three-wall structure with an open front, ventilation and temperature control become vital factors. It's also a good idea to consider sun, shade, and wind, just as with an open shelter.

Most sheds and other buildings for the backyard that you can buy or build will need to be modified for rabbits. As you consider various types of buildings, choose one made of wood. Wood not only has a solid feel, it also allows for modifications, such as adding windows, ventilation louvers, and cleanout doors, if desired. It is more difficult to make these modifications to metal sheds, which also are prone to rust and tend to simmer in summer. Vinyl sheds are another option. While they won't rust or rot, they are more difficult to work with than wood and are likely to be hot. They were never designed for animals.

Looks can mean a lot, but functionality rules. For any small backyard building, I like a fairly steep pitch on a gable roof. This helps shed rain and snow faster, and it keeps a small building from looking like a squat shack. It also provides plenty of room for vents and windows on gable ends, and a deep roof overhang keeps wind and rain away from the walls. Wide fascia boards, at least 6" or 8", help to solidify a small stance.

Additional vents or louvers should be located near the floor, next to a fan if necessary during hot weather or if a prevailing breeze does not occur, to draw air down and push it out. That will help keep your rabbits healthy. With any type of structure, the need for adequate ventilation cannot be overemphasized. Poor ventilation can lead to serious respiratory problems for your herd. As you consider various types of buildings, their adaptability for meeting your rabbit's ventilation needs should be your primary consideration.

Convenience and maintenance are other important factors. Make sure your shed either has a door wide enough to allow manure pans and a wheelbarrow or cart through, or that the walls behind the hutches can be opened to clean out the manure. If the building is wide enough, it's a good idea to place your hutches back-to-back in the center so they are away from the walls, giving you two aisles.

Finally, make sure the site you select is well drained. If you buy a small building, smaller than 8 × 12 feet, you can probably have it set on a foot or so of compacted gravel. Depending on how high you want the finished grade to be, you might want to excavate the soil 6" or so before laying the gravel. A gravel base is good for drainage and it helps to limit shifting from frost heaves in cold climates. If you place the building over gravel, it's important that the lowest framing members are pressure-treated or they sit well above the ground atop railroad ties, landscape timbers, or concrete blocks or pavers, to prevent rot. Of course, you may opt for a concrete slab with footings below the frost line. Before you decide on that, however, I'd make sure

Accommodating an Expanding Population

Five years after I built the fourth rabbitry, we moved across town. My wife thought we needed a bigger house for our growing family. I thought we needed a bigger lot, preferably with a wooded, secluded backyard. I wanted to raise more rabbits — lots more.

In rabbitry number five, the lot was about three-quarters of an acre, in a neighborhood of similar-size properties. About half of the backyards were hidden under huge oak and beech trees. Shaded, secluded, and fairly level, it seemed the perfect spot. Oh, yes, the house was okay, too.

I ordered metal shed components from a Long Island company that let you buy individual wall and roof sections to put together a shed to your own specifications. The company delivered components for two wall and roof sections, each 20 feet long, 5 feet high, and 3 feet deep. I installed them directly on the ground, 3 feet apart, and built a central roof section covered with white, corrugated fiberglass panels of the sort commonly used to shelter patios at that time. I added a screened door on each end and painted the whole structure forest green.

I also set up an automatic watering system made of plastic pipe. It had a float tank and was fed by a garden hose. I kept the system from freezing in winter with two electric cables running inside the pipe. These were controlled by separate thermostats: one turned on when the temperature dropped to freezing; the other was needed only when the thermometer fell toward zero, which hardly ever happened, even though this was northern New Jersey.

During the months those oak and beech trees bore leaves, you could hardly tell the shed was there, and the neighbors never mentioned it during the five years it existed. Was it legal? Did I need a building permit? Were 36 wire hutches housing rabbits allowed by the local authority? To this day, I don't know the answer to those questions; I never asked.

As excellent as it was, this setup fell short of perfection. The cages hung tight to the back walls, and even though I had coated the walls with a rust-preventing paint, they began to corrode. After a couple of years I had to cover the walls with exterior-grade plywood because rust had eaten holes in them. Rabbit cages next to metal walls, I learned, are not a good idea. In fact, cages next to any kind of wall can give trouble because a mixture of hay, straw, shed fur, and sprayed urine (courtesy of the bucks) creates a bad situation. I had to remove the cages and clean back there a few times a year. That's not easy when they're connected to a rigid-pipe watering system. I'll discuss improvements to this kind of arrangement elsewhere in the book.

During those five years the rabbitry produced a lot of breeding stock sold through the mail and shipped by air. That shed supplied lots of show rabbits that I exhibited at up to 10 shows per year throughout New England and the middle Atlantic states, as well as at national American Rabbit Breeders Association conventions elsewhere.

that a "permanent" foundation would not adversely impact your property tax assessment. If I were building a new structure from scratch or a kit, I'd choose a pole building design because it's inexpensive and works well with or without any kind of floor.

My Dream Barn

Because I used to be a newspaper reporter, and therefore am pretty darned nosy, I have spent a lot of time snooping around other rabbit operations, interviewing the breeders and picking up tips. I have also done some carpentry, site preparation, and foundation and electrical work. Everything I had learned went into my seventh and final rabbitry, the barn.

As I describe in *Storey's Guide to Raising Rabbits*, the barn features pole construction and was adapted from the plans for the attached garage I already had on my house. It has the same overall interior dimensions (24 × 24 feet), but I added to the barn a 4-foot roof overhang on the left for firewood storage. The floor is bank-run

All I had learned building my previous rabbitries went into constructing my seventh and final rabbitry, my dream barn.

DREAM BARN PLANS

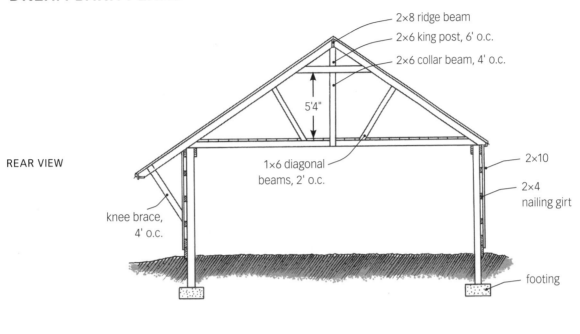

REAR VIEW

2×8 ridge beam

2×6 king post, 6' o.c.

2×6 collar beam, 4' o.c.

5'4"

1×6 diagonal beams, 2' o.c.

2×10

2×4 nailing girt

knee brace, 4' o.c.

footing

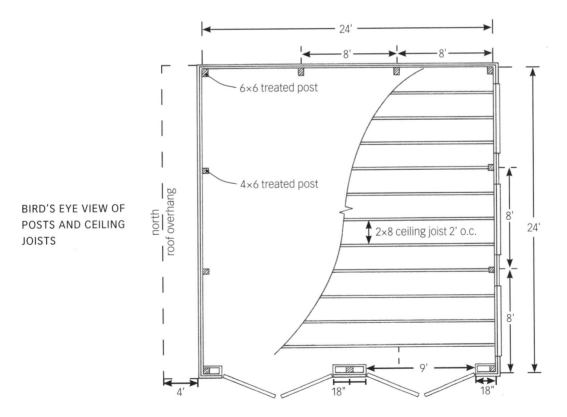

BIRD'S EYE VIEW OF POSTS AND CEILING JOISTS

6×6 treated post

4×6 treated post

north roof overhang

2×8 ceiling joist 2' o.c.

24'

8'

8'

8'

8'

8'

24'

9'

4'

18"

18"

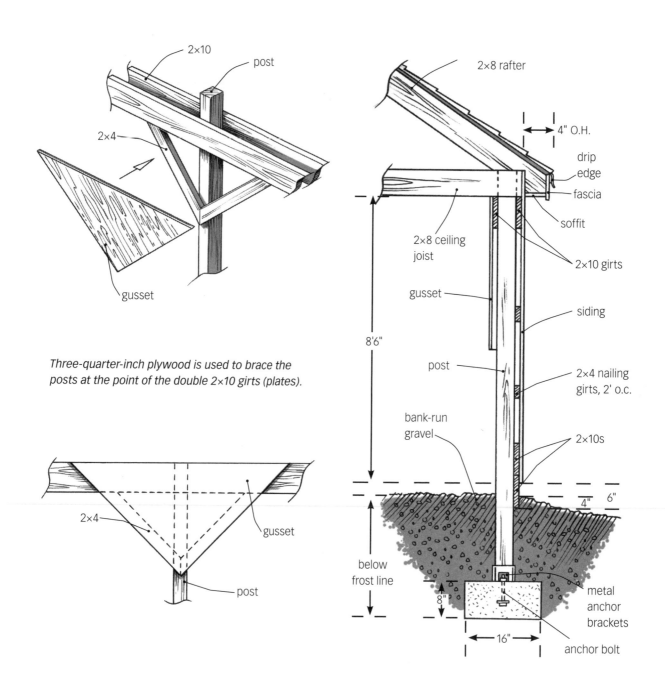

2×10

post

2×4

gusset

Three-quarter-inch plywood is used to brace the posts at the point of the double 2×10 girts (plates).

2×4

gusset

post

2×8 rafter

4" O.H.

drip edge

fascia

soffit

2×10 girts

2×8 ceiling joist

gusset

siding

post

2×4 nailing girts, 2' o.c.

bank-run gravel

2×10s

8'6"

below frost line

4"   6"

8"

metal anchor brackets

16"

anchor bolt

This profile shows the girts, siding, joists, and rafters.

gravel. The posts are on 8-foot centers and are anchored to concrete footings set below the frost line. Double 2×10 girts tie the posts together at the bottom. Double 2×10 girts support the ceiling joists and rafters. The joists are 2×8s spaced every 2 feet. Three of them are nailed to 2×6 king posts, and I installed collar beams every 4 feet. These form a great attic area while giving the barn plenty of structural strength.

I nailed asphalt shingles that match my house and garage to 1×5 hemlock roof decking. If I built it again, I might use plywood on the roof, as the 1×5s required a lot of work. I would also use pressure-treated plywood around the bottom perimeter instead of the double 2×10 skirt boards that I soaked in creosote; that plywood was not available in my area when I built the barn.

The rough-sawn, vertical pine board-and-batten siding give it the "barn" look I sought, and I set off the whole thing with semielliptical doors. On the south side I put in three 5-foot-wide windows for lots of ventilation and light. A few years later, I added an 8-foot greenhouse that spans a third of the south side.

Inside are 36 hutches that hang from joists, and there is ample aisle space along the walls. The manure falls, unimpeded by posts or hutch legs, to the gravel floor, which provides good drainage. I still like the barn the way it is; the only change I've made since I built it in 1982 was to add the greenhouse.

Shed Options

Shown on the next two pages is a variety of backyard buildings that can be modified to accommodate a small rabbitry and can also be expanded to hold as many hutches as you wish. Style is pretty much up to you, and there is a lot to choose from. My own taste is for buildings that mimic the architectural style of the existing house and garage, to keep a unified look to the property. Roofing shingles, siding, and trim that match the house and garage go a long way toward that end. You might also choose a structure that looks like a tool shed, or even a pool house or playhouse or gazebo, none of which suggest a rabbitry.

If you buy a building from a home center or a company that specializes in outbuildings, look for solid construction and quality materials. If you want to build your own and need advice on basic construction, I suggest you consult the books listed in the resources. Here you will see how various designs can be adapted to the needs of you and your rabbits, not how to build a shed from scratch. If you do decide to build your own, you will find that pole construction, which also means building with square posts on concrete footings, is a comparatively easy and inexpensive method.

❋ OUTBUILDING IDEAS FOR A RABBITRY ❋

Here are some styles of outbuildings you may want to consider for your rabbitry. See Resources for companies that will deliver buildings either completely built or that you can construct yourself from precut sections. Some may require modifications for extra ventilation and manure removal.

Playhouse or guesthouse?
Neither; it's a rabbit house.

The open section offers a storage area for feed, hay, and tools. The rabbits are inside.

The double doors allow manure to be removed in a wheelbarrow or garden cart.

The doors on the end and the front allow plenty of ventilation.

Big Space for the Big Ones — *in Michigan*

Brink's Bunnies
Dan Brink

| **LOCATION:** Rockford, Michigan | **GOAL:** Raise show rabbits |
| **AVERAGE RABBIT POPULATION:** 60–70 | **BREEDS:** Giant Chinchilla, Flemish Giant |

FOR MOST OF THE PAST 60 YEARS, Dan Brink has been raising big rabbits in big spaces, but the Rockford, Michigan, man started out small. Well, he was small.

Dan obtained his first rabbit before he started school. By age nine he was in the rabbit business, butchering his first litter and selling them dressed. He also sold live does to hospitals for $2.50 each, and then he bought does from other youngsters for 50 cents and sold them to the hospitals.

"At Easter I set up a stand in the front yard and sold a rabbit and a bag of feed for $5 with a complete guarantee that when the feed was gone or at any time I would take the rabbit back. Most of the rabbits came back and I sold them again."

Over the six decades he has been raising rabbits, Dan has rebuilt and remodeled his rabbitries several times. Like many others, he started with outside hutches. Then he built what he thought was "the perfect rabbit barn," with 32 hutches, but it wasn't perfectly ventilated. So, about 30 years after he got that first rabbit, he bought five acres so that he could build a better rabbitry.

"I built a 40 × 20-foot barn and left the back open," said Dan. "Then I added a 40 × 14-foot open-sided shed on the south side. I started out with 60 hutches and modified it to 48."

In 2010 he tore it all out and started over and ended up with 30 hutches ranging in size from 4 × 4½ feet for bucks and 4½ × 6 feet for does with litters. You read those sizes right! Dan raises Giant Chinchillas and the even bigger Flemish Giants, and they need a lot of room. When he decided he needed even more hutches, he built eight that are 4 × 7 feet each. He even has 30 more hutches outside that he uses only in the summer (see photo on page 81).

The large hutches are 2½ and 3 feet high. "It takes extra-large cages to raise extra-large rabbits with the heavy bone they need to carry their frame," he pointed out. "I use one-gallon plastic water bowls, so the rabbits always have water in front of them, and crocks for feed, so that if a rabbit goes off feed I know it immediately."

Dan finds that his big rabbits need a partially solid floor bedded with wood shavings and straw. Two feet at the rear of each floor are wire, and the droppings go out there.

His first wife, who loved to eat rabbit and had dozens of ways to cook them, passed away, and Dan remarried. "After a year of helping care for and interacting with our rabbits, my new wife, Debbie, really began to form a relationship with them. She just plain loves them, and so they no longer get eaten here," he said. "We have evolved into only raising show rabbits. We now go to lots more shows than I did before, and last year we had several 'Best of Breed' in both the Giant Chins and the Fawn Flemish. We even took three 'Best of Show' with the Flemish Giants."

Installing Hutches

When determining how to install the hutches in your shed or lean-to, you have to decide how you want to dispose of manure as well as how you will perform general cleaning and maintenance operations. An all-wire hutch is pretty light, so even with a doe and litter inside it may not weigh more than 50 or 60 pounds. That means you can suspend hutches from ceiling joists or rafters if you wish, or place them on legs.

Manure Management

First let's consider manure disposal. The two basic methods are to let the manure fall below the hutches to the floor or to catch the waste in pans.

With a gravel or concrete floor you can let manure fall to the floor directly from the bottom tier of hutches. From upper tiers, manure falls onto sloped dropping boards made of my favorite material: corrugated fiberglass panels (the kind you see on patio roofs). The manure rolls down the panels and onto the floor. With this setup, a hinged cleanout opening behind the hutches as described on pages 80 and 89 can be very handy. The simple technique I use to install dropping boards is explained on page 79.

Dropping boards that let the manure fall are preferable to pans or trays for a couple of reasons. First, you have to remove the pans to dump them almost every day because rabbits don't distribute droppings evenly over them; they usually make a pile in one corner, and the pans are only a couple of inches deep. Taking out unwieldy pans, which are as large as the

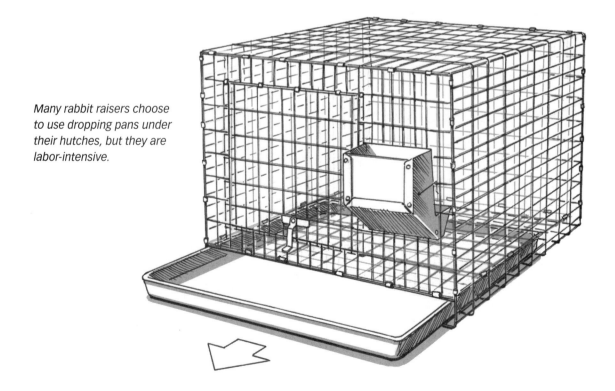

Many rabbit raisers choose to use dropping pans under their hutches, but they are labor-intensive.

hutch floor area, is not my favorite task, and I haven't even mentioned the results of the occasional spill. Second, when pans are right under the hutch floor, rabbits are constantly breathing the ammonia fumes from urine, which can lead to respiratory ailments. To combat this problem, you can put shavings in the pans or buy "disposable tray liners," which are ¼"-thick (6.4 mm) cellulose pads that soak up urine. They come in rolls and precut sheets. I don't want to pay for liners, and I don't want to deal with them, so I don't use dropping pans.

If you end up using pans with suspended hutches, you'll want the setup where the pans slide in and out right under the hutch floor. To accommodate the pan, the floor wire is installed 2" up from the bottom of the hutch sides, leaving a still-adequate ceiling height of 16" in a standard 18"-tall hutch. You can stack these hutches one on top of another, if desired, and use either metal or plastic pans (plastic pans cost more). Be sure to place the lowest hutch at a convenient height.

If you don't want to suspend the hutches, you can set them on a table or counter, or you can use leg kits from a rabbitry supply company. KW Cages is one company that offers a leg base kit for the hutch on the bottom (see Resources).

Suspending Hutches

My favorite way to install hutches is to suspend them from ceiling joists or rafters using 14-gauge wire. The wire is sold by the coil at hardware stores, home centers, and the same farm supply stores where you buy your feed. I simply attach wires at the four corners of the hutch with S-hooks that allow me to remove the hutch for periodic hosing and disinfecting outdoors. This method is a cinch with one tier of hutches. Don't worry about them swinging, because when you put two or more side-by-side they tend to stay put.

To place another tier below, I hook wires from the top of the lower hutch to the floor of the one above, leaving about a foot of space between the hutches. Then, I place a 1×6 board on edge on top of the lower hutch and attach it to the suspension wires at the ends of the row with U staples. On top of the board I place the corrugated fiberglass, which is easy to cut to fit with tin snips. I like a good 6" overhang in the front, which eliminates the need for covers on my self-feeders. I also like to have an overhang at the rear, plus several inches or even a foot between the hutches and the walls. That prevents hay, urine, and shedding fur from becoming stuck between the hutch and the wall and creating an unclean, unpleasant, and even unhealthy situation.

When modifying your shed, regardless of how you install your hutches, be sure to protect the walls. One way is to cover them with poly (polyethylene) plastic sheeting, which you can staple to the walls. You might have to replace this every year or two. A more permanent way is to cover the walls with cement board, fiberglass, or vinyl panels.

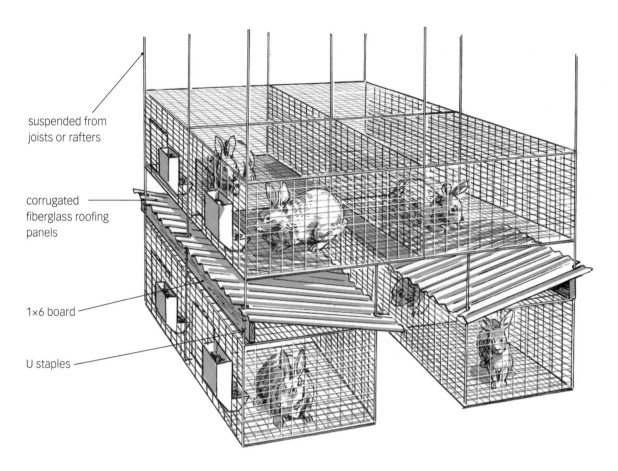

suspended from
joists or rafters

corrugated
fiberglass roofing
panels

1×6 board

U staples

Hanging hutches back to back provides plenty of access. Note also that leaving space between hutches on the lower tier allows clearance for droppings from the boards above.

Installing Hutches on Legs

The common alternative to suspending hutches is to support them on some kind of legs. Leg or stand kits offer a handy way to do this. A well-known brand of leg kit is the Stack-A-Hutch, available from Bass Equipment Company (see Resources), and similar leg kits are available from other suppliers.

The Stack-A-Hutch consists of a set of steel legs (which are powder-coated to resist corrosion) and tray slides for one hutch. You can buy the kits in various sizes to fit different hutches. They also nest into each other and can be stacked up to three high with 16"- or 18"-tall hutches or

up to four high with 14" hutches for small and dwarf breeds. You can also get casters for the legs to use on a smooth floor. The only drawback, as far as I'm concerned, is that you have to use pans with a leg kit. Nevertheless, they are very popular.

You can also make your own leg assemblies with perforated angle iron (see illustration on page 78), lumber, or even PVC plastic plumbing pipe (sold with all sorts of connecting elbows and other fittings at home centers). Another option is to buy or build shelving or racks for supporting hutches with pans. Prefab steel shelving units, previously known as steel industrial shelving, are now commonly available for

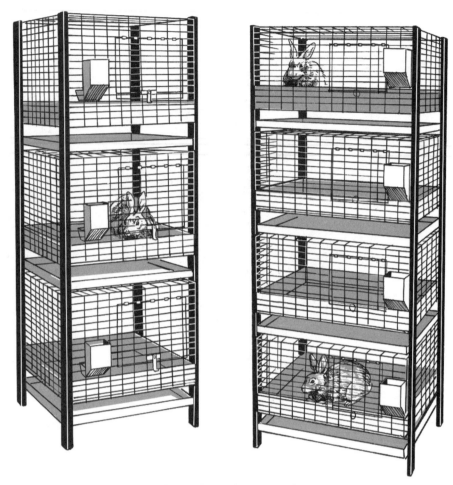

Stack-A-Hutch leg kits permit three tiers of 18-inch-high hutches and even four tiers of 14-inch-high hutches. The pans are sold separately.

consumers at home centers. Quite often you can find used industrial shelving for sale in the classified ads of your local newspaper or online classified websites. If you ask around at area warehouses, you might find some shelving that the shipping/receiving people would let you have for very little.

That said, the leg kits are a great option because they're not very expensive and can save you a lot of time, especially when it comes to cleaning hutches, since they disassemble quickly and easily. Another nice accessory available with the Stack-A-Hutch kit is the Flush-Kleen tray, a plastic pan with a corner outlet that allows you to hose down the pan and drain the contents into a pipe extension to a disposal container. These are for buildings that don't freeze; they wouldn't work if plugged with ice. For large rabbitries, you can purchase a Flush-Kleen system (see box at right) from Bass Equipment Company as well as a similar Rabbitech system from KW Cages.

✳ FLUSH-KLEEN SYSTEM ✳

The Flush-Kleen system from Bass Equipment Company lets you use a garden hose to flush away manure to a pail for removal.

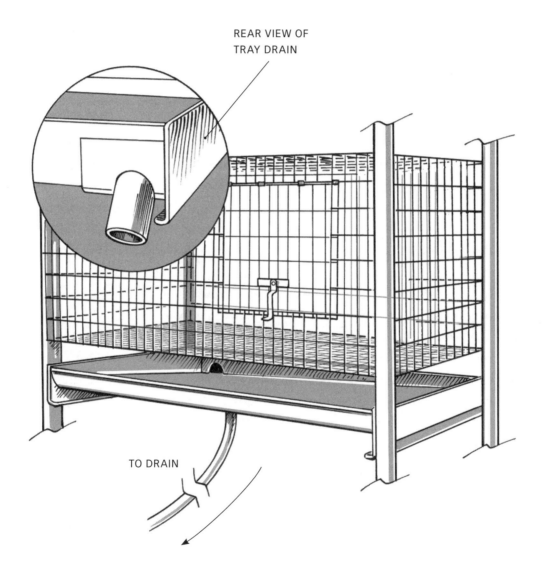

REAR VIEW OF
TRAY DRAIN

TO DRAIN

Water Tank and Lighting

When installing your hutches, be sure to leave at least 18" of headroom for a water tank, even if you are not installing a watering system right away. It's also a good idea to install some lights. More light will make your rabbitry more productive. Every day in the rabbitry should be June 21, but because that's not likely to happen any time soon, you can use fluorescent tube, CFL (compact fluorescent lamp), and even LED bulbs on a timer to stretch the length of the day to 16 hours and keep your does producing. I use the full-spectrum grow light bulbs and place a bank of them over my hutches. The greater number of litters will cover the increased electricity costs.

Fencing All Around

Just about anywhere you live, a fence around your rabbit shed is essential, unless your entire yard is already fenced in or your rabbitry sits inside a garden fence. Just because your rabbits are in a building doesn't mean they are completely safe.

A fence can play a variety of important roles, the most obvious being security. The presence of intruders can be so scary to rabbits that they can race frantically around their hutches and break their backs or actually die of heart failure due to fright. Humans can be a problem, too. Curious children can scare your stock. What's worse, there are increasingly prevalent reports of people "liberating" domestic animals. So-called animal rescue groups have claimed responsibility for some of these actions.

A good fence can also block strong winds, stop drifting snow, and provide shade.

Choosing the Fence

Consider your priorities when determining the best type and height for a fence. Do you want it to be just high enough to keep out small animals and neighborhood dogs and cats? Do you want it to complement your home's architecture and garden? Is privacy a consideration? Write down your main objectives and choose a fence that best meets these needs.

You might erect a wooden or vinyl fence that's to your liking, one that complements the style of your home, or you might prefer a wire mesh fence. Chain link would be my first choice, but 2" × 4" welded wire is a good, cheaper alternative. One kind of fence I wouldn't use is chicken wire, often called poultry netting. It may keep a wild rabbit out of your garden, but dogs can rip right through it. I wouldn't depend on it for a fence, and I would never use it on a rabbit cage. It just doesn't have the strength required for safety. You might, however, apply some tightly woven chicken wire or hardware cloth to the lower perimeter of a strong fence to keep out small intruders like rats and snakes. For added protection, a 12"-high section at the top that leans out can help keep out cats and other climbing animals.

Various kinds of vegetation can benefit your structure, from increasing the strength of the barrier to beautifying your rabbitry. A wire fence, for example, might be adorned with plantings such as pole beans. They can grow as high as 15 feet with support and thus provide a privacy screen, some shade, and, of course, some good eating. Scarlet runner beans produce bright red flowers. Other good climbing plants are morning glories, which grow rapidly, and any squash, cucumber, or gourd vine.

Ornamental Barriers

Some plantings can create an additional barrier for your rabbitry, such as a black-berry bush with its long canes. A line of shrub roses, complete with thorns, pro-vides an excellent deterrent for would-be intruders. Depending on your interest and your climate, you can choose hedge plants

Shrubs, vines, and flowers can greatly improve the look of your rabbitry.

that can grow higher than your building. The lilacs on the south side of my barn, for example, not only block the wind in winter but also shade the barn windows in the summer. By planting a hedge along your fence, especially an open wire-mesh type, you can gain most of the advantages of a solid wood fence or even a stone wall.

Trelliswork on a building or fence can boast vines or roses. I have Canadian climber roses, which don't really climb but can be tied to a trellis and trained to reach great heights. These are hardy specimens that prove you can grow roses anywhere. What's more, roses on a rabbit building are better than a neon sign on Main Street if you want to sell rabbits, because rabbit manure makes them grow. I heard years ago that rose growers (rosarians, they call 'em) are gaga over rabbit manure, so I sold them some through a local garden center. Another benefit of rabbit manure is that it doesn't need to age before it goes into the garden.

The sky's the limit for gardening possibilities. One wall of your shed can be the backdrop for a hedged herb or vegetable garden, or you could completely surround your shed with annuals and perennials, maybe even add a grape arbor. It need not be expensive — simple cedar posts can support one — although you could make one that's elaborate if you wish. Locating your shed at the end of a path lined with shrubs makes it a wonderful destination, or for more seclusion, hide it among trees, shrubs, and flowers at the end of a secret garden.

The Other Payoffs

In addition to keeping out intruders, nice fences and beautiful plantings pay off by improving the appearance of your rabbitry. If you integrate your building into your garden and landscape, instead of creating a ramshackle, smelly eyesore, you will have a thing of beauty, a gorgeous sight to behold. All who see it will admire it, including your neighbors, who will be happy to see your property value (and theirs) increase.

A beautiful appearance may also silence possible critics and win you customers. Remember that you don't *keep* rabbits the same way that you *keep* chickens. You breed them and cannot possibly keep them all even if you'd like to. Displaying your animals in attractive and functional surroundings will garner much more interest from potential customers than a rabbitry like the one my mother criticized on Mendon Mountain. This is especially true if you sell breeding stock or pets. And even if you raise rabbits only for family consumption, an inviting rabbitry makes the rabbits more palatable to people whose hearts rule their heads.

A Milk House Becomes a Rabbit House — *in Vermont*

Silver Ridge Rabbitry
Sarah Ouellette

LOCATION: Northern Vermont

AVERAGE RABBIT POPULATION: 40–60

GOAL: Meat rabbits, breeding, and show stock business

BREED: New Zealand White

SIX YEARS AGO Sarah Ouellette began to raise New Zealand White rabbits in her garage in northern Vermont.

"We just wanted to raise them for our own personal use to feed our family," Sarah told me, "but we soon found that other people were searching for breeding stock. I searched out great purebred rabbits from a meat breeding program at an agricultural school. I now have show rabbits, as well, from a top breeder."

It didn't take long for Sarah to realize that the garage just would not do the job. She is from a dairy farming family and was able to have a milk house moved from her father's property to her yard. A carpenter separated the milk house from the dairy barn, and it was loaded onto a truck and trailer. After a 30-mile trip, it was set upon a footing-and-concrete-block foundation in back of Sarah's home. The roof had to be taken off for the trip and reattached by the carpenter. At 20 × 14 feet, it was quite a load.

The milk house has windows near the ceiling on all four sides, an exhaust fan, and electricity. It's equipped with fluorescent lighting and electric heat. "I use the heat very sparingly," she told me.

"I have 28 holes [hutches]. I use 36" × 24" stacking cages with trays for my breeding stock and 30" × 24" and 24" × 24" cages for the young, growing stock (see photo on page 81). I hope to run water to the milk house next summer, as well as put up siding and paint the building. Because of the cold weather we didn't get a chance to paint it this fall.

"The garage was not an ideal situation, but now we are really cooking with the stacking cages in the milk house. I cannot say enough about setting up your rabbitry once and doing it right the first time. I sold off my old cages that didn't fit in the milk house. Now I have a new and wonderful challenge with my rabbitry. I cannot stress enough how much you want to think about your setup — the hutches, feeders, watering system, etc. It's as important as it is to find the best stock to begin with."

As Sarah's experience shows, sometimes rabbits will take you beyond your initial ideas and you have to shift gears. Sarah now takes orders for breeding stock and has a very nice website (see Resources). You will find plenty of pictures of her rabbitry there, including videos. She also has hutches and other equipment for sale that you can order to put together a rabbitry just like hers. You will have to supply your own milk house, though (or a reasonable facsimile).

CHAPTER FIVE

Building and Outfitting a Big Operation

I f you're like most people contemplating a large commercial rabbitry, you will have arrived at this juncture after raising rabbits successfully on a smaller scale. Of course, if you don't already have a suitable building for the enterprise, you'll need to build one or have one built for you.

The particulars for building one yourself are beyond the scope of this book, so before you tackle the job, see the Resources section for a list of good books that have plans, material lists, and everything else you need to know. Home centers offer additional plan books, some of which are free at the contractor's desk. Following are some of the other major considerations for setting up and operating a large rabbitry.

Heating and Cooling

You will need a large open-span building that's appropriate for your climate. Rabbitries in mild climates often can be open, unheated structures. Elsewhere, a temperature-controlled building is necessary. The size of the building depends, of course, on how many rabbits you want to raise and also how you plan to install the hutches. For years, a single tier of hutches was considered the best way to go, primarily because of the problem of manure handling, but now some double-deck hutch systems can be set up for efficient manure disposal. Two tiers instead of one can reduce your construction and heating costs but can potentially increase ventilation expense due to greater animal density. But the bottom line is your total energy bill. Feed is energy, fuel is energy, and climate and insulation influence energy costs. Because you intend to profit from the enterprise, your energy expenditures can make you or break you.

In the north, a building that is heated in whole or in part should be your goal, although you can get away with not adding heat if you have a heated watering system. When considering the cost of heat, remember that calories are units of heat. If you try to save money by not heating a building, those savings could be offset by having to feed your animals more. The cost of energy, whether in the form of artificial heat or animal feed, is the main reason that the warmer states are home to so many large commercial livestock operations. For raising rabbits, the ideal temperature is around 50°F. If the mercury pushes toward 90°F, you will need to provide cooling. Temperatures above 90°F approach the rabbits' body temperature and can kill them.

It's especially important to keep bucks cool. If the temperature exceeds 85°F, you run the risk of sterility, usually for a short time, but it can cut into your summer and early fall production and wipe out your profits. Foggers can help cool a building, and sprinkling a roof can lower the temperature inside. Insulation can be a very significant factor in both hot and cold seasons. You also don't want direct sunlight shining on your rabbits in the summer, so if that is a problem, you will do well to plant deciduous trees and shrubs for shade.

Ventilation

Regardless of the type of facility you put together, ventilation is more important than heating and cooling. If your rabbits are in wire hutches and their manure falls underneath, ideally you will bring fresh air into your building from high up and exhaust it near the floor, because you do not want to pull up ammonia fumes from below.

Precisely how much air should be moved depends on the size of your building and the density of the rabbit population. You must provide fresh air to the animals and people inside, remove excess moisture from animals' breathing and excreting urine, and prevent drafts. Insulation, air inlets and outlets, and fans are all important tools of good ventilation. If you will be housing a lot of rabbits, it's a good idea to work with an agricultural building company and its livestock building engineer and to consult with a university agricultural extension agent. These experts can advise you on the many variables related to energy and building ventilation, such as how many air exchanges you need to provide in winter and summer in your climate.

Large-Scale Rabbit Houses

You can purchase open-span buildings of various sizes from manufacturers, or you can build your own using any number of plans. See Resources for books that cover the subject.

This traditional-style building is an example of vertical board-and-batten sheathing that may be applied using local green lumber.

Large-Scale Rabbit Houses, continued

A building that provides a large unobstructed floor area for hutch installation and is tall enough for good ventilation, such as this one, makes for an ideal rabbitry space.

You can purchase steel buildings of various sizes from manufacturers.

Manure Management

In older single-tier systems, a popular method of manure handling was modeled after systems used in dairy barns. They included a poured concrete floor that sloped to a gutter and a gasoline engine outdoors that pulled a scraper blade on a chain, moving the manure out through the back of the building.

Today, at least two rabbitry supply companies offer cage-and-rack systems that eliminate scraping or shoveling manure and dumping pans. After years of field research and development, Bass Equipment Company developed the Flush-Kleen system, a completely integrated cage and cleanout system that minimizes labor and maximizes efficiency. Bass, which in the past operated a large commercial rabbitry in the Ozarks, originally designed

it with the commercial breeder in mind, although it has been accepted on a smaller scale in fanciers' rabbitries and research facilities.

The freestanding, modular design of the system makes it suitable for a variety of building situations. One-piece plastic trays contain the waste. This means it can be used over wood or concrete floors or even floors on a second story. The integrated design allows the waste to be washed down to a single drain and out to a sanitary sewer or other means of disposal. The trays fit together end-to-end, creating a seal, and bolt together with stainless steel bolts in pre-drilled holes. All of the trays come pre-drilled and marked for assembly. An end drain attaches to the end of each row in the same fashion and is fitted with a piece of 3" thin-wall PVC plastic drain pipe. From

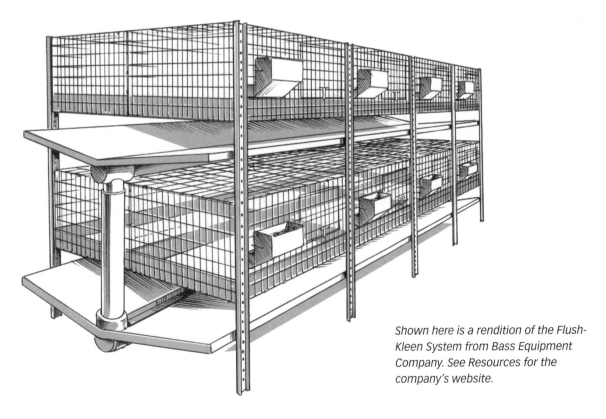

Shown here is a rendition of the Flush-Kleen System from Bass Equipment Company. See Resources for the company's website.

The Rabbitech System can be configured in rows in a large rabbitry building.

there, you drain it to a pail, your sewer or a manure holding area outdoors.

The Flush-Kleen system utilizes the standard Bass cage legs, connected by special spacers. The outer edge of the tray sits in each leg and has a standard tray slide tab. The upper deck tray rests on the lower cage. The lower tray rests on a special cross brace connected between the legs. All legs are assembled to the same length and installed on a floor that has a slight pitch to facilitate proper drainage. If you have a level floor, you can purchase an optional leg extension kit to provide the necessary slope for drainage.

The hutches for the system are constructed using baby saver wire and 14-gauge floor wire. C-rings are provided for assembly instead of J-clips because they are thicker and resist corrosion longer. These hutches are assembled as back-to-back units (they share a back wall) rather than as two individual cages. Each hutch is 24" deep, so the whole unit is 48" deep. You can vary the widths of the hutches from 12" to 26", and they are 16" high.

Bass also carries a variety of single and double-decker hutch configurations with other cleanout systems and offers consulting and custom designs.

KW Cages offers a different wash-down system called Rabbitech. It features a powder-coated H-frame with adjustable leveling, a steeply inclined polyvinyl washout with a spill lip, and a one-piece molded drain, plus the hutches, which are 30" deep, making for a 60"-deep double-sided system. You can purchase the Rabbitech system as either a double- or a single-sided system with a full plastic backsplash. KW will also help you select a system tailored to meet your requirements.

A "Temporary Rabbitry"

We moved back to Vermont because the folks at my publisher, the predecessor to Storey Publishing, asked me to come and help them sell books. I had sold a lot of rabbit books in feed stores — many more per store than the publisher was selling in bookstores. So, up we went. We bought 10 acres of land zoned for residential–agricultural use and built a house on it. You could have any animals you wanted here. There were horses across the street and cows around the corner. Today, 35 years later, there are still horses across the way and llamas and alpacas down the road, plus a few chickens, geese, and ducks.

Moving a couple of dozen breeders from New Jersey to Vermont in a Ford sedan was a challenge, but after a few trips I managed to move them all safely. You should have seen the stares I got on the highway with the empty wire hutches tied to the roof of my car. One gas station attendant along the way asked what the cages were for, and I told them they were for snakes. Where were they? The vipers must have slipped away through the mesh openings.

When the house was finished and we had moved in, I had to slap up a rabbitry in a hurry, because I had a lot of other things to do, like installing all of the interior woodwork, completing the landscaping, and building the cabinets and bookcases. So rabbitry number six was a temporary arrangement until I had time to build a barn. It was an 8 × 16-foot enclosure made from 6-foot-high stockade fencing sections with a corrugated metal roof raised to 7 feet on one long side. The roof's pitch let rain and snow slide off, and the 1-foot opening provided light and ventilation. One 8-foot fence section served as the door. I stapled sheet polyethylene to the inside walls to keep out the weather and protect the fencing. The rabbits lived in two tiers of wire hutches hung from the 2×4 roof rafters.

This "temporary" rabbitry was used for four years. All the while I pondered and planned my pole barn that became rabbitry number seven (see page 92), the one I still use today.

Inside Urban Rabbits — *in the Owner's Own Words*

Urban Rabbits **Bonny Wagoner**	**LOCATION:** Columbia City, Oregon **AVERAGE RABBIT POPULATION:** 25–35	**GOAL:** To raise and show nationally competitive rabbits **BREED:** Tan

BONNY WAGONER has at least two passions: Tan rabbits and motorcycles. She raises a lot of rabbits at Urban Rabbits, located in Columbia City, Oregon, up the road a few miles from Portland. When she's not raising rabbits or riding her bike with her husband Ric and daughter Meg, she edits and produces the Tan Tribune, the newsletter for members of the American Tan Rabbit Breeders Association. I've seen a lot of rabbit newsletters, and Bonny's is doubtless the best of them. Do I think so because the Tan is my favorite breed and years ago I was the newsletter editor myself?

How could that be?

I asked Bonny to describe her rabbitry because I know she produces some great Tans, and I wanted to know where they came from. Here's what the retired U.S. Air Force logistics officer told me, in her own words:

"Although our rabbits were originally housed outside the garage along one wall, we had to build a shed specifically for the rabbits because our neighbor would turn out her large dog and allow it to run without supervision. The dog took great sport in terrorizing our rabbits and the chicken project my daughter was working on for 4-H.

"We built a chicken coop with an enclosed sundeck and installed an electric fence around it. For the rabbits, we built a shed where they originally were housed on the side of the garage, under a walnut tree for shade in the summer. The walls were insulated and sheetrocked to enable us to wash them down during our semiannual cleaning.

Screened ventilation spaces were placed at the top of the walls on both ends. We have linoleum on the floor, which we mop as part of our routine cleaning chores. We also use one wall of our garage to house a row of cages.

"Our cages are 'condos' built as four-high units. Most are 24" × 24", but we also have some 18" × 24" cages that we use for newly weaned youngsters and for a handful of retired, elderly rabbits. We also have 24" × 36" cages that we use for the does and litters. We still have most of the original 12-year-old condos in service.

"We began with water bottles, a very labor-intensive method if you have more than a handful of rabbits. Eventually we migrated to a gravity-fed watering system by Edstrom. One of the downsides to this system was that if the water ran out of the bucket during the course of the day, all of the rabbits were out of water, so we decided to upgrade the system to be able to use city water with a regulator to control the water pressure. It works pretty well, although it does have some shortcomings, such as when a rabbit plays with the nozzle and inadvertently pulls it off. But overall, maintenance of the system is easy. The lines are flushed once a month. Most of the nozzles (drinking valves) have been in service for 10 years.

"We are very fortunate in the Pacific Northwest — we don't typically get much severe cold weather in our area. When we do get a blast of arctic cold, we use a propane heater to maintain the shed at 38°F. Because of our proximity to the Pacific Ocean and the Columbia River, we don't get much extreme hot weather, either. We use fans in the summer months for cooling, air circulation, and exhaust.

"We use trays to catch droppings, and we change our cleaning cycle based on temperature. In the middle of winter when temperatures are around freezing, we clean our cage trays inside every three days. In the heat of summer, we clean the trays every day. Otherwise, we clean them every other day. Does and litters are cleaned twice a day. Our diligence has paid off. We don't medicate our rabbits because we don't have to. Health problems with our rabbits are very unusual, and none is respiratory-related. As a result of the standards we set for ourselves we can do something many breeders won't: we guarantee the health of our rabbits when they are sold.

"Our rabbits are fed first thing each morning. All of our feeders are inside the cages, and this has proven very beneficial to our herd management. We do not use self-feeders for this reason: As each rabbit is fed, I run my hand across its back. This is an excellent way to detect early any problems such as fur block or something that is off in the rabbit, especially in our older Tans. As a side note, I think it's especially important to notice how my older Tans fare, paying particular attention to their weaker points. The Tan is one of the most stable breeds and should be shown beyond seven months, but to do this breeders must do a better job of selective breeding.

"For most of the year, I feed an 18 percent protein feed, dropping back to 16 percent over the summer. Every other day each rabbit gets a handful of hay. In the spring we hay every day during molting. Does and litters get hay twice a day. Newly weaned youngsters get hay every day. Every evening I give a small treat to each rabbit. Sometimes it's Calf Manna; sometimes it's whole oats. Depending on what's growing at the time, sometimes it's a fresh pansy or nasturtium or a clipping of blackberry vines. It's hard to find carrots with tops anymore, but when I do that also makes a fine treat.

"All of the rabbits have toys for entertainment, such as Wiffle balls, practice golf balls, and real golf balls. Being the ultimate interactive breed, Tans love to have obstacles to hop over and boxes to hop in and out of. Baby teething rings are also a hit. I'll hang a series of them on their doors, and they spend hours working to disengage each link. They also love having branches to chew on.

"We've tried a number of products of droppings tray litter, trying to find the one with the best absorption and odor control. We currently use a fir shaving of sorts that is similar to pine but not as expensive. It's more like coarse sawdust. Although it doesn't absorb liquid as efficiently as wood pellets, it does a good job controlling odor. All liquid and solid waste is removed at the same time. Rabbit waste is also known as 'master gardeners' gold.' It's one of the easiest fertilizers to use and produces phenomenal results. It gives the word *supersize* a whole new and healthy meaning."

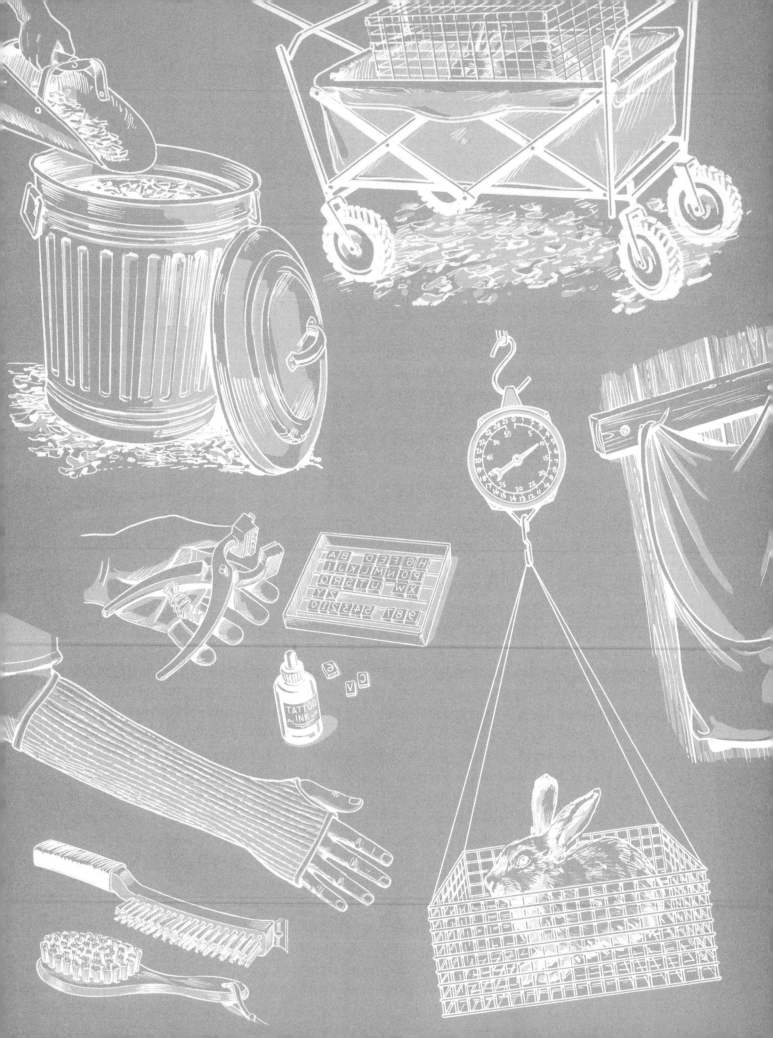

CHAPTER SIX

Tools, Equipment, and Routines

Your rabbits are in their hutches. The hutches are under cover. Now is a good time to assess what you need, and what you might simply want, to keep everything running smoothly. Once you have everything you need, it's a good idea to establish a regular routine.

The Necessities

Let's start with you — with what to wear — and then proceed to what your rabbits need.

LONG-SLEEVED TOP

After handling rabbits you realize that while they have a reputation for being soft and cuddly, they are also noted for scratching and shedding. When handling them, long sleeves are a must. Wooly sweaters are not a good idea, as rabbits' nails can get caught in them, and shedding hair sticks to them. A twill or nylon shop coat, a full canvas apron over a sturdy shirt or blouse, or a sweatshirt make a good outfit for the rabbitry. I have a favorite thigh-length denim jacket.

ROSE GLOVES OR BOOT SOCK

When carefully looking over young stock at weaning time, deciding what to keep and what to cull, you realize what sharp nails they have. And they tend to aim for the insides of your wrists and forearms. My favorite antidote is a pair of rose gardener gloves with the fingers snipped off. These are leather gloves with very long, heavy leather gauntlets that protect up to your elbow. Rose gardeners use them because

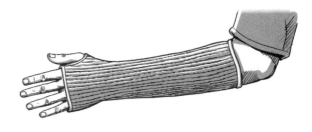

You can make arm protectors out of heavy boot socks.

they are always reaching deep into thorny bushes to do some pruning. I first used these gloves many years ago after I took a physical. The physician looked at my wrists and forearms and asked, "Are you and the missus getting along?"

A good alternative to these gloves is an old pair of high, heavy boot socks. If you cut the toes off and poke holes for your thumbs, you can pull them right up over your shirt sleeves and handle your rabbits with safety. I have observed rabbit judges wearing athletic sweatbands on their wrists. Those are good, and perhaps stylish, but they don't provide as much coverage as the rose gardener gloves or the modified boot socks.

FEED SCOOP AND BUCKET

You will need a feed bucket, unless you have a large operation and need a feed cart. A 20- or 30-gallon steel garbage can will safeguard your feed supply. Plastic feed scoops are inexpensive, and you can easily make one from a plastic bottle. I have a steel scoop that is still as good as new after 40 years of use every single day.

NAIL CLIPPERS

Rabbits raised on wire tend to wear down their nails over time, so clipping them is not a regular chore. But when you need to clip the nails of an older animal, a regular dog nail clipper does the job nicely. You don't want a rabbit's nails to get caught in the floor wire of a cage.

BRUSHES

A good wire brush with a scraper on the end (sold in paint and hardware departments) comes in very handy for use on hutch floors to remove the occasional dried manure droppings that cling to the

You will need a pail or bucket that will securely store your feed.

A wire brush (top) is an indispensable tool in the rabbitry. Also useful are a brush with nylon bristles (middle) and a slicker brush (bottom).

wire. A wire brush with brass or stainless steel bristles will cost more but last longer. Another brush with nylon bristles is handy for brushing hair from the hutch. And a slicker brush, as used on dogs and cats, will take out dead fur and ease a rabbit through a molt.

BUILDING TOOLS

We covered the tools that you need to build your hutches in chapter 2. They include wire cutters, J-clip and hog ring (or C-ring) pliers, and basic household tools.

BUCKET AND SCRUB BRUSH

These are for washing and disinfecting hutch floors and feeders, as well as nest boxes, unless you are using wire nest boxes with disposable corrugated cardboard liners.

PROPANE TORCH

Not required but certainly handy is a small propane torch to burn off hair on wire hutches. Use a torch nozzle that flares, rather than pinpoints, the flame. My Tans don't shed a whole lot, but in late summer or early autumn, my torch goes to work.

WIDE-BLADE PUTTY KNIFE OR PAINT SCRAPER

If you use pans under the hutches, these are useful for scraping off manure that the hose can't budge.

PITCHFORK AND SHOVEL

Use for manure removal. A rake and hoe are handy, too.

Other handy items for the rabbitry include a propane torch to burn hair from hutches and an aluminum photo reflector to provide heat under nest boxes when temperatures plummet.

WHEELBARROW OR GARDEN CART

For hauling manure to the garden or compost heap. I have one of those wooden garden carts with big bicycle wheels, and it still works great after 30 years of service.

NEST BOX WARMER PAD OR ALUMINUM PHOTOFLOOD REFLECTOR

If your rabbitry is cold in winter, you will need one of these, along with a 25-watt light bulb or a heat bulb used to keep pet reptiles warm. See page 57 for more information.

INSULATION FOAM BOARD PANELS AND PLASTIC FOAM TRAYS

These are another way to make doe hutches warmer, along with the heating devices above.

Optional Items

The following equipment and supplies are useful in many rabbitries, but others may be required depending on the kind of operation you have. For example, if you raise Angoras, you may need a grooming table and tools such as combs and shears.

HANGING SCALE

This is perfect for weighing fryers and other rabbits to make sure they are reaching their growth goals. The hanging type has a hook to hold a bucket or other container and subtracts the weight of the container. I like to weigh my rabbits in one of my wire nest boxes. You can purchase the scales in capacities of either 22 or 55 pounds.

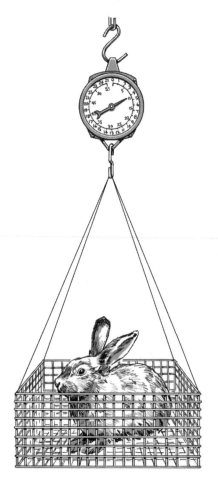

The all-wire nest box can come in handy when weighing your rabbits.

LIMESTONE OR DEODORIZER

A bag of agricultural limestone, the granulated kind used on lawns and gardens, is handy to have in hot weather. You can sprinkle some on manure under hutches or in pans to diminish ammonia odor, if necessary. Neutralizing ammonia odors will help prevent respiratory problems. Additionally, you can purchase biodegradable deodorizers that eliminate odors in a building. One product, called Anotec (available from Bass Equipment Company; see Resources), chemically bonds with odors to eliminate them, not just mask them. You can apply it directly to surfaces or use it as an air freshener in a pump dispenser.

RECORD SHEETS

Record keeping will make you a proficient rabbit raiser. You need hutch cards to record mating dates and the does' overall performance. Stud cards for each buck let you measure the performance of your bucks at a glance. These cards fit on the outside of hopper feeders and are held on with clear plastic card covers. There is so much plastic in packaging these days that you can probably make your own covers, but otherwise you can buy them in sizes to fit the feeders. Pedigree forms, production record sheets, and daily, monthly, and annual summary sheets also are available from rabbit supply companies, or you can devise your own. In addition, now there are computer programs that help you keep track of everything to make sure you have a successful operation.

SKINNING KNIFE, SKINNING HOOKS, AND HIDE STRETCHER

These are needed only if you are slaughtering your own meat rabbits. A good knife is a "boning knife," which is used in most commercial slaughter plants because of its ideal size and shape for rabbit processing. Skinning hooks, made of plated steel, are specifically shaped to hold the carcass in the correct position for dressing. Mine are attached to a board that I hang on the wall of the barn. I like to place a large black plastic garbage bag under everything to catch the offal.

TATTOO SET

If you plan to enter your rabbits in shows, they need an identifying number or letters in the left ear. And even if you don't show, after you raise a few rabbits you will find no better way to keep track of who's who in the herd. I prefer the pliers type, but you can also buy one that is battery-operated. If you don't have room for a table in your rabbitry, you can make a shelf that folds down from the wall and use that for tattooing.

I like to hang a plastic garbage bag on my skinning hooks to contain offal when I butcher a rabbit.

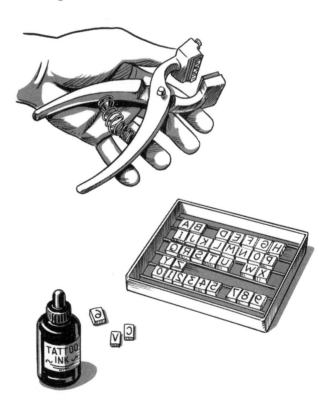

Use a plier-type tattoo tool to identify your stock with numbers and letters.

Miscellaneous Equipment

Depending on your location, you may need equipment for such needs as heating, cooling, or lighting, among others. Here are some items I have found useful.

COOLING TOOLS

A portable electric fan can come in handy in hot weather. You can even find one that you hitch to a garden hose to mist your rabbitry with cool, moist air. A soaker hose, which emits a fine spray, can be placed on your shed roof. It cools the roof, and when the water flows off, any breeze blowing by can provide additional cooling. If temperatures in your area are regularly in the 90s or above, you can make your rabbits (especially your pregnant does) more comfortable by filling some plastic milk jugs or soda bottles with water and freezing them.

If you put a couple of them in their hutch, they will lie against them. I keep herd bucks in my lower tier of hutches because it's cooler near the floor. You might also purchase a window air conditioner. The classified ads often reveal many serviceable used ones at very reasonable prices.

SMALL ELECTRIC HEATER

I keep one in my barn, not for my rabbits but to warm my fingers if I'm knocking ice out of water pans. There are many different kinds of electric heaters, and you might want to use one on occasion to warm up your rabbitry.

FOLDING CART OR TWO-WHEEL TRUCK

If you go to rabbit shows, this will help you haul carrying cages from the parking lot to the showroom. If you raise Angoras, you'll also want a folding grooming table

A folding cart that can easily fit in your vehicle comes in handy when you are transporting rabbits to a showroom for judging.

to take to the show, along with assorted combs, brushes, shears, and clippers.

HUMANE TRAP

Mine is known as the Havahart. It traps any rabbits that happen to escape their hutches. I just bait it with a carrot or some lettuce, and I'll have the rabbit back in its hutch by the next day.

MOTION-ACTIVATED OUTDOOR LIGHTS

These can help protect your rabbits from nighttime intruders and can also be used along walks and on stairs for your safety. You can even buy one that is battery-operated or solar-powered if you don't have electricity in your shelter. These lights turn on when the sensor detects movement and shut off when no more motion is detected.

CLIP AND SOLAR LIGHTS

If it's dark in your backyard when you feed your rabbits, holding a flashlight at the same time is not easy. A battery-powered light that clips to your glasses or the bill of a cap frees up your hands. You can also buy solar lights with long power cables that allow you to place the solar panel unit outdoors in the sun and locate the lamp inside where you need it. This kind of light can be especially useful to boost chances of conception when days are short. Solar walkway lights leading up to the rabbitry provide for safe passage, and installing them takes nothing more than poking each fixture into the ground.

A motion-activated light can be useful for safety and protection.

If you need to feed your rabbits during darkness, a lighted cap such as this will leave your hands free.

Rabbitry Routine

Throughout this book I have placed a lot of emphasis on security when constructing or purchasing your building and choosing your equipment. Once you have a safe rabbitry in place, security should still be a priority.

This security extends beyond protecting rabbits from outside threats and includes their safekeeping in all ways, from vigilant temperature control to regular feeding schedules. The best way to make sure your rabbits stay protected is to develop consistent routines.

Visitor Guidelines

A rabbit raiser called me up recently and said his does would not take care of their litters, and one of his bucks mysteriously died, and did I have any idea why. It took a few questions, but I finally learned that three cats resided in his rabbit shed. Cats and rabbits are natural enemies. Cats eat rabbits when they can catch them, and if they can't, they can still worry rabbits so much that they race around frantically in their hutches until they die of fright or, probably more accurately, heart failure. The does can also become so worried that they neglect their maternal duties and scatter and even eat their babies.

I wouldn't let a cat or a dog, not even a nice friendly one, into my rabbit barn. The rabbits might not recognize its friendliness, and the animal might be carrying tapeworms that could be passed on to the rabbits. An infected dog or cat can leave behind worm eggs in their feces and, if the eggs are consumed by a rabbit, these will form cysts in the intestines and under the skin. Dogs and cats should not be in the rabbitry or near any feed, utensils, or nest box bedding such as straw or shavings.

I wouldn't let boisterous children into my rabbitry either, because they too might worry the rabbits. I wouldn't even let in other adults who are rabbit raisers unless I knew they ran a disease-free rabbitry, because I don't want my herd to be infected. Visitors who do not have rabbits are welcome, of course, but you should first find out why they want to be there. There are increasing reports of people from animal "rescue" groups letting rabbits out of hutches. If you obtain a new rabbit from another raiser, quarantine it in a hutch in your garage or outdoors for a couple of weeks before you let it join the rabbitry.

You should also eliminate rats and mice, which can eat feed and may carry disease. These rodents will often urinate on stored feed bags or in the hopper feeders. In addition, they leave their droppings all over the place and may get into hutches and kill baby rabbits or frighten the does to the point that they will kill their litters.

If you house chickens or other livestock in your building, it's a good idea to partition the rabbit area to keep them out. Chickens love to hop up on rabbit hutches and deposit their droppings. Goats, sheep, pigs, cows, and horses should have their own space specially designed for them. Rabbits are easily startled, and a surprise can be especially detrimental when a doe is kindling. You want to keep the environment calm and quiet — no grunting, mooing, crowing, or honking. What's good for the goose may be good for the gander but *not* for the rabbit.

When I'm not around, I keep a padlock on my barn door, because I don't want anyone going in there without my knowledge.

Feeding and Breeding Routines

Security for my rabbits also means a regular feeding schedule. It doesn't really matter whether you feed your rabbits once or twice a day, but you should do it at the same time every day, because they will come to expect it. If I enter the barn at feeding time, they are glad to see me and are active but calm. If I go in at another time, I greet them by saying, "Hi, gang." Otherwise, they are nervous until they can see me.

I also like a regular routine. I like to mate my rabbits on weekends so that youngsters will be born at midweek 31 days later, when neighborhood children are in school and things are quiet. This also means that I put in the nest box during a weekend, which is useful because I am less likely to forget to check their hutch cards for their due dates during the weekend. I like to palpate and test mate at two weeks, again on the weekend.

My routine includes placing weanlings into their own hutches — does together for a few weeks and bucks in separate quarters. At this time, I tattoo their ears and enter their personal information in my

stock record book. I record their ear number, date of birth, sex, the numbers of the sire and dam, and any special remarks, such as the quality of their fur or body type. The pertinent information also goes onto a new card for each hutch.

Good Housekeeping

Now that you've put together an excellent rabbitry, complete with all-wire hutches, you'll want it to continue functioning at a high level. Overall cleanliness will do a lot toward making your life and those of your rabbits a lot more pleasant. The lack of it can lead to disease and be a turnoff to visitors, including potential customers. You, the rabbits, your family, neighbors, and customers will appreciate your diligence in cleaning, even if they don't come right out and thank you. It's the finishing touch to a rabbitry and will help you raise a healthy and productive herd.

Before taking the weanlings from their mother, I clean and disinfect their new homes. First I clean with a wire brush and burn off hair if necessary, making sure not to keep the torch on the wire longer than necessary to remove the hair, so it won't damage the galvanizing and ultimately rust the cage. The torch cleans the cage but does not disinfect it, so I follow with a spray disinfectant. You can use a commercial disinfectant according to the product directions, or make your own by mixing an ounce of household bleach in a quart of water. A bottle sprayer works fine for a small operation, while a tank sprayer is best if you have a lot of hutches. I like to let the hutch dry all day after I spray the disinfectant, then I spray plain water and let that dry before the rabbits go in.

While you are washing, disinfecting, and rinsing the hutch, don't forget to do the same to the nest boxes, feeders, and pans. You can spray them or dip them in a bucket of disinfectant, then rinse and dry them in the sun, if possible. If you use the all-wire nest boxes and give them a new corrugated cardboard liner for each new litter, you will save yourself some work and time and will always have a germ-free box.

Regularly clean off hair, cobwebs, and dirt from the tops of cages, as well as suspension wires, legs, and any other supporting members. Dust can lead to respiratory problems. Brush hutch floors every day with a wire brush to get rid of any manure that has not fallen through. Make the brushing a part of your daily routine. It won't take long, it will save you a lot of hassle in the long term, and you are not likely to forget if you make it a daily occurrence. Keeping your cages clean and dry will help avoid sore hocks, too. If the rabbits have wet hair on their feet, the hair doesn't shield the skin, and sores can develop.

Because my barn is well ventilated and sits on more than a foot of gravel with good drainage, the manure under the hutches stays pretty dry. Wet manure attracts flies and creates ammonia fumes, so it should be removed regularly and not left to sit for long. Because the manure stays dry under my hutches, I let it stay there until I need it for the garden or the compost heap, which is conveniently located behind the barn. I don't move manure around in hot weather but instead give it a good sprinkling of granulated agricultural limestone. This neutralizes the alkalinity of the ammonia and makes it a better candidate for the compost. If urine stains develop on a concrete or wood floor, a couple of ounces of vinegar in a bucket of water should take care of them.

A small greenhouse on the south side of my rabbit barn helps keep the fly population way down in the summer. When the plants are in the ground outdoors and the greenhouse is empty, flies tend to congregate in there. At the same time, spiders weave webs in it and love to catch and kill flies, so I call it my "website."

Downsizing in the Animal World

Moxie Meadow Rabbitry
Barbara Gaul

LOCATION: Buffalo, New York
AVERAGE RABBIT POPULATION: 20

GOAL: Breed to Standard and Best of Breed at National Show
BREED: French Angora

BARBARA GAUL, a nurse who lives near Buffalo, New York, had a good time breeding, raising, training, and showing Arabian horses. She also raises layer chickens and turkeys. Then, a back injury, age, and a major car accident put an end to raising large animals. Still longing to have animals around, she realized that she could work with a small animal that took up less space but nevertheless involved regular hands-on care. After lots of reading and conversations with rabbit breeders and her county 4-H leader, she eventually settled on French Angoras.

Why French Angoras? "They have a sweet disposition, apparently enjoy human company, and of course are a dual-purpose rabbit," Barbara said. "Their wool is highly desirable, and with a growing interest in handspinning, is a marketable product.

"Grooming is not difficult. You can find most of the equipment in the drugstore or the pet store. You do really need a grooming table, but they are easy to make with simple tools and basic skills. The tabletop should be about 2 feet square. The top should be at about waist height. Grooming, which includes combing and blowing with a small electric blower, is fun, easy, and it must be done regularly. Just lay a folded bath towel on the tabletop — so the rabbit can get a grip with its toe-nails — to groom the top of the rabbit. It's easy to do the underside and trim nails if you put the towel on your lap and gently position the rabbit on its back."

Grooming is fun, it's easy, and it must be done regularly.

Barbara is adamant about the need for wire cages. She says that good air circulation is especially important with Angoras for overall health and so you can see all your rabbits easily. She clips them for comfort at the beginning of summer, reminding me that rabbit wool is nine times warmer than sheep wool. Two oscillating tower fans and a window fan cool the rabbits when needed.

"Housing the rabbits in an attached garage helps keep the rabbits cool," she said. "Also, they are close at hand for all care. It's easy to keep them safe. Washtubs are nearby for frequent washing of all the equipment." Then she added, "The washtubs are also for washing your hands." Spoken like a nurse.

Resources

RABBIT EQUIPMENT SUPPLIERS

Check out your local feed or farm supply store. Stores across the nation carry welded wire mesh, J-clips, C-rings or hog rings, clip or ring pliers, feeders, water bottles, prefabricated hutches, and more. Most will order what you need if it's not in stock, as their distributors carry extensive lines of rabbit equipment.

Bass Equipment Company
Monett, Missouri
800-798-0150
www.bassequipment.com
Free color catalog includes just about anything a rabbit raiser needs. Formerly operated a commercial rabbitry. Also does consulting. In business since 1961.

FarmTek
Dyersville, Iowa
800-327-6835
www.farmtek.com
Galvanized wire mesh, tools, hardware, buildings, heating, cooling, lighting, insulation, fencing, and just about anything else you can possibly need to build a rabbitry.

KW Cages
Santee, California
800-447-2243
www.kwcages.com
Carries a complete array of rabbit hutches and equipment. Maker of the Rabbitech cage system and provides consultation. Owner is an experienced rabbit raiser.

SMALL BUILDINGS

The Carriage Shed
White River Junction, Vermont
800-441-6057
www.carriageshed.com
Large variety of outbuildings suitable for modification for rabbits

Jamaica Cottage Shop
Jamaica, Vermont
866-297-3760
www.jamaicacottageshop.com
Fine garden buildings that can be adapted for rabbitries

LARGE BUILDINGS

Lester Building Systems, LLC
Lester Prairie, Minnesota
800-826-4439
www.lesterbuildings.com
Longtime supplier of big barns and other farm buildings

Morton Buildings, Inc.
Morton, Illinois
800-447-7436
www.mortonbuildings.com
Excellent large structures for livestock

130

BOOKS

Bennett, Bob. *Storey's Guide to Raising Rabbits*, 4th ed. Storey Publishing, 2009.
Rabbit raising how-to, including information on housing and equipment.

Burch, Monte. *Building Small Barns, Sheds & Shelters.* Storey Publishing, 1983.
Contains plans for my rabbit barn and several other excellent buildings suitable for rabbit housing.

Damerow, Gail. *Fences for Pasture & Garden.* Storey Publishing, 1992.
Advice and complete instructions for building fences.

Ekarius, Carol. *How to Build Animal Housing.* Storey Publishing, 2004.
Contains plans for sheltering rabbits.

Seddon, Leigh W. *Low-Cost Green Lumber Construction.* Garden Way, 1981.
Contains wonderful, economical barn plan with detailed, stepwise building instructions, excellent illustrations, and materials list.

———. *Practical Pole Building Construction: With Plans for Barns, Cabins & Outbuildings.* Williamson Publishing, 1985.
Excellent plans with illustrations and instructions. Includes a Lean-to Animal Shelter, ideal for a small rabbitry.

Storey, John and Martha Storey. *Storey's Basic Country Skills: A Practical Guide to Self-Reliance.* Storey Publishing, 1999.
Covers the complete range of things to know if you live in the country, including excellent plans for a tool shed that would make a fine small rabbit building in town.

Vivian, John. *Building Fences of Wood, Stone, Metal, & Plants.* Williamson Publishing, 1987.
Complete information with excellent stepwise instructions and both photographs and line drawings of just about every conceivable kind of fence, plus hedges and other landscaping.

Wolfe, Ralph. *Low-Cost Pole Building Construction.* Storey Publishing, 1980.
Classic book on pole building, an ideal method for constructing a rabbit building, large or small.

WEBSITES OF RAISERS PROFILED

Sarah Ouelette
Silver Ridge Rabbitry
www.silverridgerabbitry.com

Callene and Eric Rapp
Rare Hare Barn, LLC
www.rareharebarn.com

Bonny Wagoner
Urban Rabbits
www.urbanrabbits.net

OTHER RESOURCE

American Rabbit Breeders Association, Inc.
P. O. Box 426
Bloomington, IL 61702
www.arba.net

Index

Page references in *italics* indicate illustrations or photographs.

A

all-wire hutches. *See also* plans, all-wire hutches
  basics, 15–16
  as best option, 7, *7*
  door, 19, *19*
  fasteners for, 18
  multiple, *30,* 30–31
  plastic flooring for, 20, *20*
  prefabricated, assembly of, 17, *17*
  sizing of, 18–19
  in tour of rabbitries, 81–88, *81–88*
  urine guards, 29, *29*
  welded wire, 16, 18
American Chinchilla, 83, *83*
American White, 82, *82*
angle iron framework, 77–78, *78–79*
Angora, *86,* 86–87, *87,* 124–25, 129
ARBA (American Rabbit Breeders Association), 66–67, 91, 131
awnings/shades/tarps, 9, 74, *74, 75,* 77, *77*

B

baby saver/baby saver wire, 28, *28,* 35, *35*
barns. *See* closed buildings
Bennett rabbitries, vi, 31, 53, 65, 91, *92–94,* 92–95, 115
Blackberry Farm, 12–13
Blanc de Hotot, 82, *82*
breeding routines, 127
Brink, Dan, 81, *81,* 98
Brink's Bunnies, 98
Broken New Zealand, 84–85, *85*
brushes, *120,* 120–21
building codes, 71

building environment, 22
buildings, prefabricated, 130
building tools, 121

C

Californian, 38–39
carrying cage, *64,* 64–65
cart, *124,* 124–25
Champagne D'Argent, 84–85, 88, *88*
cleanliness, 127–28
closed buildings
  basics, 90, 92
    dream barn/plans, *92–94,* 92–95
    sheds, *81,* 85, 95–97, *96–97*
considerations, primary, 10–11
cooling tools/equipment, 109–10, 124
creep feeders, 45, *45*
C-rings, 18, 114

D

Daniels, Bobbi, *86,* 86–87, *87*
decoy rabbit, 11
deodorizer/limestone, 122
disinfecting hutches, 128
door, 19, *19,* 27, *27,* 34
drainage, 90
dropping boards, 79, 101, *101*
dropping pans, 99, *99*

E

easy wire hutch plan, *21,* 21–28
  assembly, 26, *26*
  baby saver, 28, *28*
  cutting/shaping wire, 23–25
  cutting the door, 27

exploded view, 25, *25*
fixing wire, 90-degree angle, 24, *24*
flattening wire mesh, 24, *24*
tools and materials, 22, *22*
wire options, 23, *23*
elements/weather. *See also* awnings/shades/
    tarps
    freezing temperatures, 49, 53–55, *54*
    as primary consideration, 11

F

fasteners, 18, *18*, 87, *87*
feeders, 41–45, *43*
    buying and using, 45, 47
    creep, 45, *45*
    hay/pellet, 42, 45
    hay racks, 44, *44*, 81, *81*
    types of, 41–42
feeding routines, 127
feed scoop and bucket, 120, *120*
fencing, 104–6
    choices in, 104–5
    landscaping and, 105, *105*, 106
    ornamental barriers, 105–6
fiber for spinning, *86*, 86–87, *87*, 129
fixing wire, 90-degree angle, 24, *24*
flattening wire mesh, 24, *24*
Flemish Giant, 98
Flush-Kleen system, 102, 103, *103*, 113, *113*, 114
folding cart, *124*, 124–25
French Angora, *86*, 86–87, *87*, 129
front "wall" options, 74, *74*, *75*, 77, *77*

G

garage lean-to, *88*, 88–89, *89*
Gaul, Barbara, 129
Giant Chinchilla, 98
Green Meadow Farms, 46

H

hangers, 22, *22*
hay/pellet feeders, 42, *43*, 45
hay racks, 44, *44*
heating equipment, 49, 53, 55, 57, *57*, 109–10,
    121, *121*, 124
Hogan, Kerry, 88, *88*

hog rings, 18, *18*, 26, *26*
Holland Lops, 12–13
hutch. *See* installing hutches; plans, all-wire
    hutches; *specific type of hutch*

I

installing hutches
    Flush-Kleen system, 102, *103*
    on legs, *101*, 101–2
    manure management and, 99–100
    suspending hutches, 100, *101*
    water tank and lighting, 104
installing watering system, 50–55
    basics, 50–51
    flexible tubing for, 51, *52*, 53
    freeze-x system, 54, *54*
    rigid pipe system, 53, 55
insulation, 55, 121

J

J-clips, 18, *18*, 26, *26*, 114
J feeder, 41, *43*

K

Krahulec, Jim, 66–67

L

landscaping, 76, *105*, 105–6
large-scale rabbitries, 109–14, *111*, *112*
    heating and cooling, 109–10
    manure management, *113*, 113–14, *114*
    ventilation, 110
latches, 22, *22*
laws and neighbors, 11, 71
lean-to
    construction of, 89
    garage, 80, *80*, 89, *89*
legs for hutches, 35, *35*
lighting, 104, 125, *125*
Lockley, R. M., 5

M

manure management
    double doors for, 97, *97*
    dropping boards, 79, 101, *101*
    dropping pans, 99, *99*

manure management (*continued*)
    greenhouse and, 128
    hutch installation and, 99–100
    for large-scale rabbitries, 113–14
    limestone/deodorizer, 122
    panel for, 80, *80*
    predators and, 9, *9*
    tools for, 121
McCune, Myke, 38–39
milk house as rabbitry, 81, *81*, 107
Mini Satins, 66–67
morants, 2, 3, *3*
Moxie Meadow Rabbitry, 129
multiple hutches, *30*, 30–31
    ten small hutches, 32–34, *33*
    top-opening hutch, *35*, 35–37, *37*

N

neighbors and laws, 11, 71
nest boxes, 56–63
    flanges vs. no flanges, 56–57
    heating, 57, *57*, 121, *121*
    plans, *58*, 58–62, *59, 60, 61, 62*
    subterranean, 63, *63*
    wood vs. metal, 56
New Zealand White/Black, 46, 107

O

Ouellette, Sarah, 81, *81*, 107, 131
outbuildings. *See* closed buildings

P

partitions, 30, *30*
plans
    angle iron shelter, *78–79*
    wire nest box with flanges, 58–60, *58–60*
    wire nest box without flanges, 61–62, *61–62*
plans, all-wire hutches
    easy, 21–28, *21–28*
    ten small, 32–34, *33*
    top-opening, *35*, 35–37, *37*
plastic flooring, 20, *20*
pole building, 72–73, 76
Polymax flooring, 20, *20*
predators, 2, 9, *9*, 126
prefabricated wire hutch, 17, *17*

Q

quonset-style hutches, 36

R

Rabbitech System, 113, 114, *114*
rabbits running free, 4–5
"rabbit tractors," 6
Rapp, Eric and Callene, *82*, 82–83, *83*, 131
Rare Hare Barn, 81, *81*, *82*, 82–83, *83*, 131
record keeping, 122–23
Rex rabbits, 20
Roper, Gabriel, 81, *81*, *84*, 84–85, *85*
routines
    feeding/breeding, 127
    housekeeping, 127–28
    visitor guidelines, 126–27

S

salt/salt spools, 49
security, 126–27
sheds. *See* closed buildings
shelter for rabbitry. *See also* angle iron frame-
      work; closed buildings; three-wall
      structures
    building/zoning requirements, 71
    creature comforts, 69, 71
    location for, 71
    simple backyard rabbitry, 70, *70*
S-hook, *35*, 37, *37*
Silver Fox, 88
Silver Ridge Rabbitry, 107, 131
size of rabbitry, 10
skinning equipment, 123, *123*
spring and S-hook, 37, *37*
Stack-A-Hutch, 101, 102, *102*
stacked hutches, 81, *81*
subterranean nest box, 63, *63*
suspending hutches, 100–101, *101*

T

Tan rabbits, 116–17
tattoo set, 123, *123*
ten small hutches plan, 32–34, *33*
    attaching latch and door, 34
    cutting the door, 32–33
    tools and materials, 32

three-wall structures, 71–89, *72, 73*
  basics, 71
  basic shelter with angle iron framework,
    77–78, *78, 79*
  front "wall" options, 74, *74, 75*
  pole building, 72–73, 76
tools/equipment
  attire, 119, *119,* 125, *125*
  cooling, 109–10, 124
  fasteners, 18, *18,* 87, *87*
  hand tools, 22, *22,* 24, *24,* 26, *26*
  heating, 49, 53, 55, 57, *57,* 109–10, 121, 124
  miscellaneous, *124,* 124–25, *125*
  necessities, 119–21, *119–121*
  optional, *122,* 122–23, *123*
  spring and S-hook, 37, *37*
  suppliers of, 130
top-opening hutch plan, *35,* 35–37, *37*
trays, *9, 9*

U

Urban Rabbits, 116–17, 131
urine guards, 8, *8,* 29, *29*

V

ventilation, 90, 97, *97,* 110
Vigue, Debbie, 12–13, 81, *81*
visitor guidelines, 126–27

W

Wagoner, Bonny, 116–17, 131
watering methods/systems, 47–55. *See also*
    installing watering system
  automatic/semiautomatic, 50, 52, *52*

crock, 47, *47*
freezing temperatures and, 49
holding tank for, 85, *85*
hutch installation and, 104
metal and hard plastic vessel, 48
tin can, 47
water bottle, 48, *48–49*
weather. *See* elements/weather
weighing rabbits, 122, *122*
welded wire, 16, 18
wide mouth feeder, *43*
Wilson, R. L., 46, 81, *81*
wire cutters, 22, *22*
wire hutch. *See* all-wire hutches
wire mesh rolls, 16, *16,* 23, *23*
wire nest box with flanges
  fitting the cardboard, 60, *60*
  flange construction, 59, *59*
  nest box plan, *58,* 58–59
  tools and materials, 58, *59*
wire nest box without flanges, *61,* 61–62, *62*
wire options, 23, *23*
wood-and-wire hutch
  described, 6, *6*
  modification of, 8, 8–9, *9*
wood protection, 8, *8*
work space, 22

Z

zoning laws, 71
Zoom Rabbitry, 66–67

Other Storey Titles You Will Enjoy

BY THE SAME AUTHOR

Storey's Guide to Raising Rabbits, by Bob Bennett.
Everything from breeding and caring for your rabbits to showing guidelines
and marketing advice.
256 pages. Paper. ISBN 978-1-60342-456-1.
Hardcover. ISBN 978-1-60342-457-8.

The Backyard Homestead Guide to Raising Farm Animals,
edited by Gail Damerow.
Expert advice on raising healthy, happy, productive farm animals.
360 pages. Paper. ISBN 978-1-60342-969-6.

Chicken Coops, by Judy Pangman.
A collection of hen hideaways to spark your imagination and inspire you
to begin building.
176 pages. Paper. ISBN 978-1-58017-627-9.
Hardcover. ISBN 978-1-58017-631-6.

How to Build Animal Housing, by Carol Ekarius.
An all-inclusive guide to building shelters that meet animals' individual
needs: barns, windbreaks, and shade structures, plus watering systems,
feeders, chutes, stanchions, and more.
272 pages. Paper. ISBN 978-1-58017-527-2.

Storey's Guide to Raising Chickens, by Gail Damerow.
The ultimate guide that includes information on training, hobby farming,
fowl first aid, and more.
448 pages. Paper. ISBN 978-1-60342-469-1.
Hardcover. ISBN 978-1-60342-470-7.

Storey's Guide to Raising Miniature Livestock, by Sue Weaver.
The ins and outs of breeding, feeding, and housing miniature goats, horses,
donkeys, and more.
464 pages. Paper. ISBN 978-1-60342-481-3.
Hardcover. ISBN 978-1-60342-482-0.

These and other books from Storey Publishing are available
wherever quality books are sold or by calling 1-800-441-5700.
Visit us at *www.storey.com*.